Text Book of
Disaster Management

Related Books

Titles	*Authors*
— 21St Century Biology and Agriculture: Textbook Series (Vol. 1-4)	*A. Raman*
— A Handbook of Environment Impact Assessment	*V.S. Kulkarni*
— A Handbook of Research Process	*A. Raman*
— Disaster Perception and Management	T.S. Chouhan
— Antriksh Ek Khoj (Hindi)	R.V. Bothale
— Applied Microbiology (Agriculture, Environmental, Food and Industrial Microbiology)	S.M. Reddy
— Biodiversity and Environmental Biotechnology	P. Dwivedi
— Biodiversity of Ranthambhore Tiger Reserve Rajasthan	V. Singh
— Climate Agriculture and Man Shrinking Biodiversity and Sustainability	V.K. Sehgal
— Climate Change & Himalayan Ecosystem: Indicator, Bio & Water Resources	J. Sundaresan
— Climate Change & Himalayan Informatics	J. Sundaresan
— Climate Change and Himalaya: Natural Hazards and Mountain Resources	J. Sundaresan
— Climate Change and Its Ecological Implications for the Western Himalaya	V.L. Chopra
— Climate Change Impact on Ecosystem	J. Sundaresan
— Ecology Workbook	R. Misra
— Remote Sensing: Principles and Applications *2nd Ed*	A.N. Patel
— Climate Change and Environment	J. Sundaresan
— Climate Change Adaptation Strategies in Agriculture and Allied Sectors	*GSLHV Prasad Rao*
— Environment At Crossoads: Challenges and Green Solutions	Arun Arya
— Environmental Pollution and Environmental Management	P. Dwivedi
— Environmental Protection	A.K. Thukral
— Essentials Of Environmental Studies **	S.S. Sagwal
— Forest Ecology **	S.S.Sagwal
— A Text Book Of Environmental Science	Vidya Thakur
— Vayu Pradushan	D.D. Ojha
— Objective Environ. Studies & Disaster Management	M. Prasanthrajan
— Objective Environmental Pollution and Climate Change	M. Prasanthrajan
— Ocean and Coastal Ecology (21St Century Biology and Agriculture:Textbook Series)	K. Kathiresan
— Pollution Risk and Impact on Social, Economic and Environmental Sectors And Remediation**	Golam Kibria
— Prayavaran Avabodh (Hindi)	D.D. Ojha
— Prayavaran Javiki: Environmental Biology(Hindi)**	M.M. Trigunayat
— Prayavaran Prabhandhan (Hindi)	D.D. Ojha
— Questions and Answers in Environmental science practical	R.K. Kamble
— Environmental Engineering and Safety	Sangeeta Raut
— Wildlife Ecology and Conservation (21St Century Biology and Agriculture: Textbook Series)	M. Balakrishnan

Text Book of Disaster Management

Ajay Kumar Srivastava
Scientist-Agrometeorology
JN Krishi Vishwa Vidyalaya, Jabalpur
Campus-College of Agriculture, Tikamgarh (M.P.)

SCIENTIFIC® PUBLISHERS

Published by:

SCIENTIFIC PUBLISHERS (INDIA)
5A, New Pali Road, P.O. Box 91
Jodhpur 342 001, India
E-mail: info@scientificpub.com
Website: www.scientificpubonline.com

Print 2025

ISBN : 978-93-89412-47-5
eISBN: 978-93-89412-46-8

Printed in India

Dedicated to...

My beloved and respectful wife

Smt. Nilu Srivastava

※

Foreword

श्री संस्थान

Society For Rural Improvement

Registered under Rajasthan Societies Act, 1958, Registration No : 556-Jaipur-2008-09

Disasters are as old as human history but the dramatic increase and the damage caused by them in the recent past have become a cause of national and international concern. The frequency and severity of local and regional disasters have already reached to a crisis. Natural disasters occur frequently in India and all over the world at irregular intervals and at unexpected moments. Their impact on society has increased rapidly in recent decades due to population pressure and man-made environmental degradation. Drought, frost, floods and other manmade disasters are already posing serious problems particularly in Madhya Pradesh, where the means of its management and mitigation measures are scares.

Now India is facing critical problems like pollution, over exploitation of natural resources, and deadly natural disasters than ever before. One of the important challenges of disaster management is to create awareness in public and to achieve this goal, Government and Non-government agencies are conducting different programmes in India. As a part of this, education institutes and universities are called upon to take appropriate actions for making people and students aware of the dangers ahead and the steps they must take to check and mitigate their ill-effects. Teaching of disaster management in schools and colleges level is the part of this ongoing awareness programme. Students are an integral part of community, and hence, they must be prepared well advanced and adequately to prevent, face and respond to disasters.

The course materials on this topic is scattered and pertain to different domain of science. Bringing these diverse materials under one compilation is indeed a good effort made by the author. I congratulate the author for taking the step forward for bringing this book and sincerely hope that it proves to be an effective and practical guide for students, scientists and other experts.

(Laxman Singh Rathore)
President

International Consultant, The World Bank
Former Director General of Meteorology & Permanent Representative of India with World Meteorological Organization

112-A, Nemisagar Colony, Vaishali Nagar, Jaipur - 302 021 • Ph. : 0141-2351526, Mob. : 9680080080
Web. : www.srisansthan.org • Email : gs@srisansthan.org

Preface

Every Indian has learned the idea to cope with distress situation from their childhood. Being a Philosophical country the idea of coping with distress situation is taught in India from century. Disasters occur when natural, technical, or human-induced phenomena impact on vulnerable socio-economic systems. India is highly prone for many disasters and many a times unable to manage properly due to lack of cooperation, awareness and management skills. The role of society and as well government is needed to reduce the development gap between rich and poor in coping with disastrous situation. In recent past Government of India has introduced the Disaster Management course in School and College to equip students with the skills of management in disasters situations. National as well as state level disaster management institutes have been opened for planning and implementation of disaster related programmes in India.

Disaster management is an applied and multidisciplinary subject. Students irrespective of their disciplines should have the introductory knowledge of disaster management. In spite of providing the class room knowledge about concepts and practices of natural disaster management for development of the students will be a distant dream unless and until their shoulder may not be responsibility for its management in disastrous and unforeseen conditions. So it's very essential to equip the people to conduct thorough assessment of losses from different hazards starting from village to national level and discoursed the management and mitigation strategies.

Over many years of teaching experience at under- graduate and post graduate level, I realized that no under-graduate level textbook is presenting the relevance and dire needs of disaster management to the students in a simple and understandable manner. I believe that until the fundamental things of a subject are explained in an interesting way, a student's tendency towards that subject cannot be established, and especially during the course of B.Sc. (Agriculture) where, a student has to study many more diverse subjects. Although the subject of disaster management is not new and related study materials are accessible through many books available in the market, but the single compilation covering diverse aspects of disaster management in special context to agriculture is still lacking.

In view of the above this book has been prepared on the topic as per new ICAR fifth Dean committee report and divided into eleven chapters. This book has addressed the key concepts and practices of disaster management and describes about natural disasters its types as well as manmade disasters have also been discussed. Different aspects of disaster management like role, responsibilities, rescue and rehabilitation are described. I hope that the material would be important not only for teaching purpose but it also equipped the students to face and overcome from disastrous situation. I hope the book will endow a good understanding of the subject "Disaster management" to the students of agricultural faculty.

I bow my head towards the Grace of God without, which it is not possible to complete this hard task.

May, 2020 **A. K. Srivastava**

Acknowledgments

This book although took few years for writing but actually it is the blessings of my lifetime gurus, who always inspire me and fuelled me for such preparation. I would like to thank my revered guru Dr. B.C. Panda; who set example in teaching which helped me to develop a thirst for knowledge as well as methods for quenching that thirst. I am greatly indebted to Prof. Vinay Sehgal, Professor, Department of Agricultural Physics, Indian Institute of Agricultural Research, New Delhi for his continuous and selfless comments on this subject. Besides, I pay my gratitude to a variety of mentors, friends and colleagues specially Dr. P.K. Singh who helped me and gave valuable suggestions for improving the text. I am particularly indebted to Dr L.S. Rathore, International Consultant, The World Bank and Former Director General of Meteorology and permanent representative of India with World Meteorological Organization, who generously agreed to write a foreword for this book. This text book is written in very simple manner and utilizated important materials from various public domain portals, books, descriptions and reports (Annexure V) is also thankfully acknowledged.

I am always indebted to my competent authorities at JNKVV, Jabalpur namely Prof. P.K. Bisen, Hon'ble Vice Chancellor, Dr. Dhirendra Khare, Dean, Faculty of Agriculture, Dr. Abhishek Shukla, Director Instruction, Dr. P.K.Mishra, Director Research Services, and Dr. A.K.Sarawagi. Dean, whose mentorship, advice and continuous guidance gave me strength and courage to continue my work. I would like to thank my family for patiently enduring the writing of this book, which seemed at times to be an endless process. Finally, I would like to take this opportunity to express my gratitude to everybody for their collective work and unselfish devotion to the publication of this book. The help and cooperation renders from friends, faculty members are dully acknowledged. My sincere appreciation also goes to Mr Tanay Sharma of Scientific Publishers, Jodhpur for their support in bringing this book in this form.

Ajay Kumar Srivastava

Contents

Acronym

AAI	:	Aridity Anomaly Index
AIR	:	All India Radio
AISMR	:	All India Summer Monsoon Rainfall
ARMVs	:	Accident Relief Medical Vans
ATI	:	Administrative Training Institute
BIS	:	Bureau of Indian Standards
BMTPC	:	Building Materials & Technology Promotion Council
BOD	:	Biochemical Demand of Oxygen
BSF	:	Border Security Force
CAP	:	Contingency Action Plan
CBSE	:	Central Board of Secondary Education
CBOs	:	Community Based Organisations
CBRN	:	Chemical, Biological, Radiological and Nuclear
CCC	:	Central Control Committee
CCMNC	:	Cabinet Committee on Management of Natural Calamities
CCS	:	Cabinet Committee on Security
CD		Civil Defence
CFCs	:	Chloro Fluoro Carbons
CG	:	Central Government
CGWB	:	Central Ground Water Board
CISF	:	Central Industrial Security Force
COBSE	:	Council of Board of School Education
COD	:	Chemical Oxygen Demand
CP	:	Commissioner of Police
CPMF	:	Central Para Military Forces
CPWD	:	Central Public Works Department
CR	:	Control Room
CRF	:	Calamity Relief Fund
CRPF	:	Central Reserve Police Force
CSR	:	Corporate Social Responsibility
CWC	:	Central Water Commission

ENSO : El Niño-Southern Oscillation
GDP : Gross domestic production
DAC : Department of Agriculture & Cooperation
DDMA : District Disaster Management Authority
DDMP : District Disaster Management Plan
DGCD : Director General Civil Defence
DM : Disaster Management
DM : District Magistrate
DMC : Disaster Management Cycle
DMI : Disaster Management Institute
DO : Dissolved Oxygen
DoRD : Department of Rural Development
DoRT&H : Department of Road Transport & Highways
DORSCON : Disease Outbreak Response System Condition
DORS : Disease outbreak response system
DOS : Department of Space
DOT : Department of Telecommunication
DRDO : Defence Research Development Organisation
DRP : Disaster Recovery Plan
DWR : Doppler Weather Radar
DWS : Department of Drinking Water Supply
ECC : Emergency Control Centre
ECC : Emergency Command Centre
EOC : Emergency Operations Center
ERC : Emergency Response Centre
ERZ : Earthquakes Risk Zone
ESF : Emergency Support Functions
EU : European Union
FACT : Field Assessment and Coordination Team
GIS : Geographic Information System
GoI : Government of India
GPS : Global Positioning System
GSI : Geological Survey of India
HFCs : Hydro Fluoro Carbons
HLC : High Level Committee
HMV : Hidden Markov Model

HPC	:	High Powered Committee
HPC	:	High Performance Computing
HPCL	:	Hindustan Petroleum Corporation Limited Refinery
HUDCO	:	Housing and Urban Development Corporation
HFA	:	Hyogo Framework for Action
IAEA	:	International Atomic Energy Agency
IAEM	:	International Association of Emergency Managers
I&B	:	Ministry of Information & Broadcasting
IATF	:	Inter-Agency Task Force
IC	:	Incident Commander
ICAR	:	Indian Council of Agricultural Research
ICMR	:	Indian Council of Medical Research
ICS	:	Incident Command System
ICT	:	Information and Communication Technology
IDNDR	:	International Decade for Natural Disaster Reduction
IDKN	:	India Disaster Knowledge Network
IDRN	:	India Disaster Resource Network
IDS	:	Integrated Defence Staff
IGNOU	:	Indira Gandhi National Open University
IFRC	:	International Federation of Red Cross and Red Crescent Societies
IITs	:	Indian Institutes of Technology
IMC	:	Inter-Ministerial Committee
IMD	:	India Meteorological Department
IMG	:	Inter-Ministerial Group
INCEDE	:	Inter-national Centre for Disaster Mitigation Engineering
INCOIS	:	National Centre for Ocean Information Services
INFFRAS	:	Indian Forest Fire Response and Assessment System
IO	:	Information Officer
IRP	:	International Recovery Platform
ISDR	:	International Strategy for Disaster Reduction
IPCL	:	Indian Petrochemicals Corporation Limited
IT	:	Information Technology
ITBP	:	Indo-Tibetan Border Police
ITIs	:	Industrial Training Institutes
ITK	:	Indigenous Technical Knowledge
MoA	:	Ministry of Agriculture

MoES	:	Ministry of Earth Science
MoHA	:	Ministry of Home Affairs
MoHRD	:	Ministry of Human Resource Development
MMI	:	Mercalli Intensity
MOD	:	Ministry of Defence
MOF	:	Ministry of Finance
MOH	:	Ministry of Health
MoPRRD	:	Ministry of Panchayati Raj & Rural Development
MOUD	:	Ministry of Urban Development
MNCFC	:	Mahalanobis National Crop Forecasting Center
MPGOs	:	Multi-Parametric Geophysical Observatories
MWR	:	Ministry of Water Resources
NCAP	:	National Contingency Action Plan
NCC	:	National Cadet Corps
NCS	:	National Center for Seismology
NCCF	:	National Calamity Contingency Fund
NCMC	:	National Crisis Management Committee
NCDM	:	National Centre for Disaster Management
NCMRW	:	National Centre for Medium Range Weather Forecasting
ND	:	Natural Disaster
NADAMS	:	National Drought Assessment and Monitoring System
NDDR	:	National Day for Disaster Reduction
NDEM	:	National Database for Emergency Management
NDMA	:	National Disaster Management Authority
NDMF	:	National Disaster Mitigation Fund
NDMS	:	National Disaster Management System
NDRF	:	National Disaster Response Force
NDVI	:	Normalized DifferenceVegetative Index
NEC	:	National Executive Committee
NGO	:	Non-Governmental Organisation
NIDM	:	National Institute of Disaster Management
NIT	:	National Institutes of Technology
NRSA	:	National Remote Sensing Agency
NSDI	:	National Spatial Data Infrastructure
NSS	:	National Service Scheme
PAHs	:	Polyacrycrylic Aromatic Hydrocarbons

PCP : Penta Chloro Phenol
PPP : Public-Private Partnership
PRIs : Panchayati Raj Institutions
R&D : Research and Development
RTSMN : Real Time Seismic Monitoring Network
SAARC : South Asian Association for Regional Cooperation
SAP : State Armed Police
SART : Search & Rescue Team
SDDM : State Department of Disaster Management
SDMA : State Disaster Management Authority
SDRF : State Disaster Response Force
SEC : State Executive Committee
SEMA : State Emergency Management Authority
SFS : State Fire Services
SG : State Government
SOP : Standard Operating Procedures
SPI : Standardized Precipitation Index
SPWD : State Public Works Department
SRI : Standardized Runoff Index
SWM : South West Monsoon
SWIR : Short Wave Infra-Red
SSART : State Search & Rescue Team
TCE : Tri Chloro Ethylene
TWS : Tsunami Warning System
UGC : University Grants Commission
ULBs : Urban Local Bodies
UN : United Nations
UNEP : United Nation Environmental Programme
UNU : United Nations University
UTs : Union Territories
VEI : Volcanic Eruption Intensity
WCDR : World Conference on Disaster Reduction
WMO : World Meteorological Organization
WSSI : World Seismic Safety Initiative

CHAPTER 1

BASICS OF DISASTER

A disaster is an unforeseen and impulsive event which causes widespread loss of life or property in any human habitat area and an outside assistance is required to cope up the situation.

1.1 Meaning and Definition of Disaster

The term 'Disaster' is made from a French word *desastre*, which is a combination of two words 'des' meaning bad and 'aster' meaning star. Thus, the term 'disaster' refers to 'Bad Star'. Disaster is considered to be an outcome or outburst of some unfavourable star or effect of a bad star in history. The word *disaster* implies a sudden overwhelming and unforeseen event.

Different definitions have been given by different authors and some of them are given below to understand the disaster and area of its impacts.

Susman *et al.* described the disaster as the interface between an extreme physical environment and a vulnerable human population. Such definitions emphasize the fact that the socio-economic and political factors are of paramount importance in understanding; why populations are vulnerable to the environment and experience disasters.

Anderson (1990) defined natural disasters as a temporary event triggered by natural hazards that overwhelm local response capacity and seriously affect the social and economic development of a region.

As per United Nations (UN) "disaster is a serious disruption of the the functioning of society, causing widespread human, material or environmental losses; which exceed the capacity of the affected society to cope using only its own resources".

According to World Health Organization (WHO) a disaster is- "Any occurrence that causes damage, ecological disruption, loss of human life, deterioration of health and health services on a scale sufficient to warrant an extraordinary response from outside the affected community". World Meteorological Organization (WMO) definition included any acts, which disruption the health services at large scale or very much affected and needs outside response to restore the health services is also a disaster.

Centre for Research on the Epidemiology of Disasters (CRED) classified an event as a disaster if at least "10 people are killed and /or 100 or more are affected and/or an appeal for international assistance is made or a state of emergency declared".

From the above definitions, a disaster may be defined as the realisation of a hazard that inversely impacts things of human values. Thus it may include:

1. A chance; an accident.
2. A chance of being injured or harmed.
3. A possible source of danger: a hazard.
4. A state of emergency.
5. A possible loss of life, property, health and environment at large scale.
6. Needs an extraordinary response from outside.
7. Devastate local response capacity and seriously affects the social and economic development.

Finally one can define – A disaster is a catastrophe, which causes widespread destruction and distress to the life and property of a human-populated area and during a disaster the persons of the affected area are not able to cope up (with available resources) without outside assistance.

> *By definition, a disaster can only occur in a constructed or built environment where human activity has taken place. In other words, a disaster cannot occur in an uninhabited area that has never been touched by humans.*

Disasters occur when natural, technical, or human-induced phenomena impact on vulnerable socio-economic systems.

Any hazard, like flood, earthquake or cyclone; which is a triggering event along with greater vulnerability (inadequate access to resources, sick and old people, lack of awareness etc.) would lead to disaster causing greater loss to life and property. For example; an earthquake in an uninhabited desert cannot be considered a disaster, no matter how strong its intensity may be produced.

1.2 Features of a disaster

A disaster is a result from the combination of hazard, vulnerability and insufficient capacity or measures to reduce the potential chances of risk. A disaster happens when a hazard impacts on the vulnerable population and causes damage, casualties and disruption. A disaster may be illustrated much better through the Figure 1.1.

Figure 1.1 : Disaster and causing factors

A disaster also has its underlying causes and triggering events:

i) Underlying causes

Limited access to resources, physical status, training/skills, disabilities, age, poverty, education etc., are the underlying causes of a disaster.

ii) Trigger event

All hazards are also known as trigger events and include the following:

Earthquake, Tsunamis, Floods, Cyclones, Volcanic eruption, Drought, Landslide, War Technological accident, Environmental pollution etc.

- *A disaster disturbs the normal functioning of society and affects people and the environment.*
- *It results in large scale destruction of life and property.*
- *It affects in such a way that only external assistance is capable enough to cope with destruction.*

1.3 Indian Scenario

India due to its unique geographical characteristics and high population density is one of the world's most 'disaster-prone' country. Every year a number of natural hazards like drought (Rajasthan or Western Odisha), floods (Ganga-Brahmaputra plain), cyclones (East Coast), landslides and earthquakes (Himalayas) etc. occur in different parts of the country in varying intensity. The socio-economic conditions of the country, further affect the extent to the damage done by these hazards.

India's Natural Disasters Proneness

The Indian subcontinent is one of the most disaster-prone areas of the world and highly prone to natural disasters like floods, droughts, frosts, cyclones, landslides, earthquakes, cloud burst. The total geographical area of the country is 329 million hectares and almost 85 per cent of the area is vulnerable to hazards. Out of the 29 states and 7 union territories, 22 are disaster-prone. These states are Rajasthan, Madhya Pradesh, Uttar Pradesh, Bihar, Jharkhand, Jammu and Kashmir (JandK), Maharashtra, Odisha, Karnataka, West Bengal, Himachal Pradesh, Tamilnadu, Andhra Pradesh, Uttarakhand and 7 sister states of north-east. Almost 60 per cent of the land is vulnerable to the earthquake (high seismic zones III–V) and 8 per cent to cyclones, 12 per cent to floods (Figure 1.2). About 68 per cent of the arable land is vulnerable to drought. India has also vulnerable to tsunamis since the occurrences of a tsunami in 2004 over the Indian Ocean.

Disaster-proneness: *It is the likelihood of a place or area being affected by any type of disaster.*

Figure 1.2 : India disaster proneness percentage : 60 % (Earthquake), 16% (Drought), 12% (Flood), 8% (Cyclone), 15% (Landslide and Avalanche).
(*Source:* Collage of online search free download image)

Social Vulnerability

- 90 percent people affected by disasters are poor, marginalized and small farmers.
- Pockets of poverty are Semi-arid/Arid regions, Flood Plains, Mountainous regions of north-east, Himalayan, etc.
- Disasters further breed disparity and social imbalances.

Economic Impact

- 4.5 to 10 per cent Loss to National GDP
- Some states (Assam, Bihar, AP, Maharashtra, Odisha)

  — Loss to the extent of 10–20 per cent of state's GDP
- Drought of 2001–02

  — Led to 15 per cent reduction in food production and significant drop in GDP (210 Mt to 180 Mt).

Disaster prone landmass of India

➤ *About 60 per cent of landmass is prone to earthquakes.*

➤ *About 12 per cent of landmass is prone to floods (40 million Hectares).*

➤ *About 8 per cent of landmass is prone to cyclones (Nearly 3 lakh sq. km.).*

➤ *About 68 per cent of the land under cultivation is prone to drought.*

Floods, droughts, cyclones, landslides and cloud burst are the recurrent phenomenon in India. The Indian coastline, spread over by 8041 kilometres is exposed to tropical cyclones. The region semi-arid/arid of north-west and the semi-arid region of central India is the site of droughts, while the Indo-Gangetic plains are of flood.

Susceptibility to disasters is compounded by frequent occurrences of manmade disasters such as fire, accidents, epidemics etc. Natural calamities and man-made devastations have affected the financial stability of the people as well as the nation.

Disaster Prone Zones

India has been divided into 5 zones according to its disaster proneness to natural disasters (Figure 1.3);

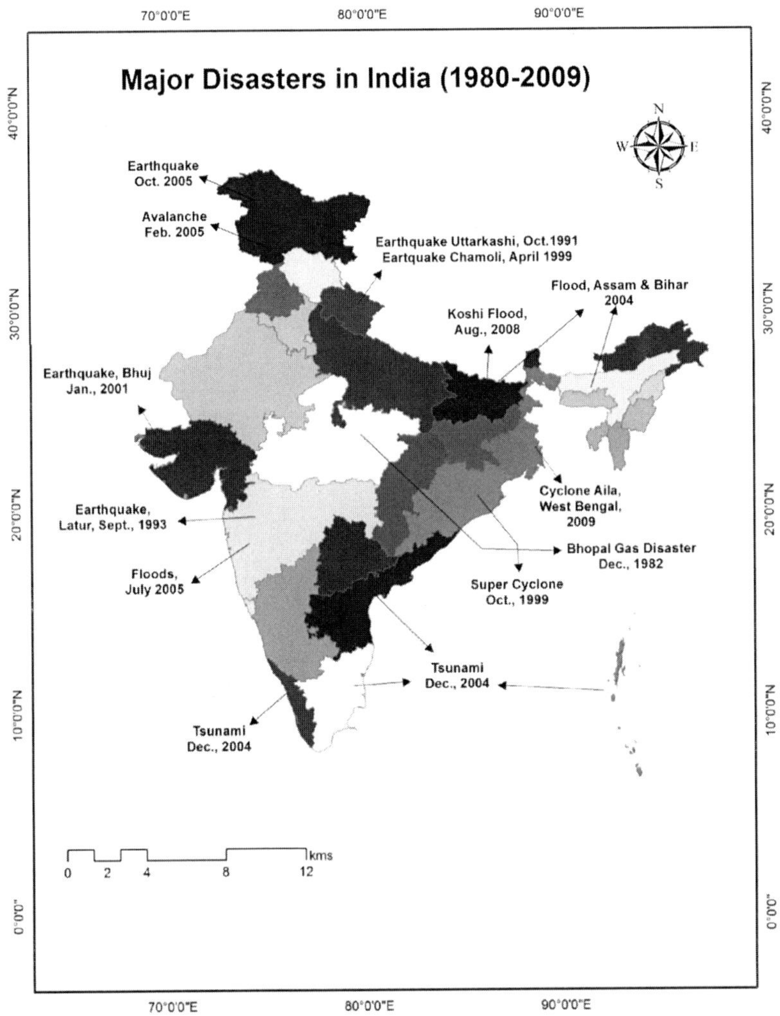

Figure 1.3 : Disaster proneness regions of India (*Source:* IMD, New Delhi)

1. **Northern mountain region including foot-hills**; this region is earthquake prone belt with violent subterranean volcanic activity and is prone to avalanches to landslides and strong cold waves
2. **Indo-Gangetic plains**; this region is prone to floods.
3. **Deccan plateau**; this region is drought-prone area.
4. **The western desert**; this region is drought-prone area.
5. **Coastal areas**; this region is prone to cyclones and tidal waves

Numerous disasters were occurred in past, but their occurrence frequency has been increased sharply in recent decades (Figure 1.4). The average number of disasters events (natural and manmade) has increased sixty per cent during 1999 to 2003; over the past decade. The average number was 428 per year from 1994 to 1998 and 707 per year from 1999 to 2003. The increase of the disasters events are much more in developing/underdeveloped countries, where it increased up to 142 per cent. In the decade 1990–2000, an average of about 4344 people lost their lives and about 30 million people were affected by disasters every year.

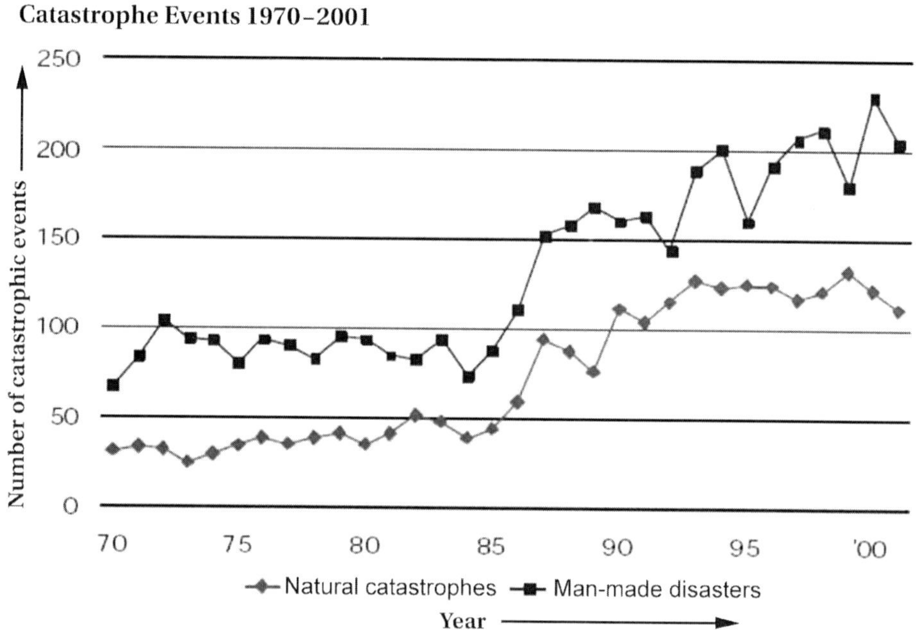

Figure 1.4 : Decadal occurrence of disasters

Each year natural disasters take the heavy loss on human life and property and number of natural disasters over the past fifty five years given in Fig.1.5. The United Nations estimates that in the past 20 years, nearly three million lives have been lost due to natural disasters and 800 million people have been affected (Katayama, 1994). Extensive damage is inflicted by natural disasters, which also kill more than one million people every decade.

Number of Natural Disasters over the Past Fifty-five years

Disaster prone landmass of India

Nearly 90 percent of losses of life due to natural disasters were caused by weather-climate and water-related hazards

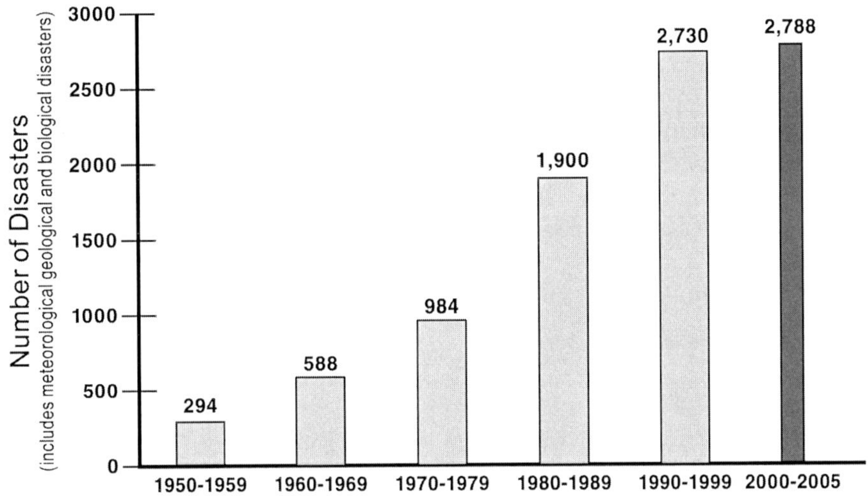

Figure 1.5 : Natural disasters over the past five and half decades (Courtesy from UN/ISDR - "Disaster Statistics" *http://www.unisdr.org/disaster-statistics/occourence-trends-century.htm*)

Many extreme disasters were occurred in India, has affected a large number of people and also results in heavy loss to livelihood and property. This closely resembles the global scenarios. The super cyclone of Odisha (1999), the Gujarat earthquake (2001) and the recent Tsunami (2004) affected millions across the country leaving behind a trail of heavy loss of life, property and livelihood. The major disasters that have caused a massive impact on the Indian community, Table 1.1 shows some of the important disaster list.

Table 1.1: Major disasters in India since 1970

S. No.	Date	Place	Disaster	Loss of life
A) Cyclone				
1.	29th October, 1971	Odisha	Cyclone and tidal waves	10,000 people dead
2.	19th November, 1977	Andhra Pradesh	Cyclone and tidal waves	20,000 people dead
3.	29th–30th October 1999	Odisha	Cyclone and tidal waves	9,000 people dead

S. No.	Date	Place	Disaster	Loss of life
4.	01st November, 2012	A.P. and T.N.	Cyclone and heavy rain	2 people dead
B) Earthquake				
5.	20th October, 1991	Uttarkashi		723 people dead
6.	30th September, 1993	Latur		8,000 people dead
7.	22nd May, 1997	Jabalpur		39 people dead
8.	29th March, 1997	Chamoli		100 people dead
9.	26th January, 2001	Bhuj		10,000 people dead
C) Landslide				
10.	July 1991	Assam		300 people killed
11.	August 1993	Nagaland		500 people killed
12.	18th August 1998	Malpa		210 people killed
D) Flood				
13.	1978	North East India		3,800 people killed
14.	1994	Assam		More than 2000 people killed

- ➤ *Between 1990 and 2000 an average of about 3400 people lost their lives annually and about 3 crore people were affected by disasters every year.*
- ➤ *About 17,000 people died by the Tsunami on 26 Dec. 2004.*
- ➤ *About 170 died, 210 injured and 70 are severely injured 20 November, 2016, Pukhraya, UP India in a rail accident.*

A wide variety of forces have been identified/responsible for the occurrence of natural disasters. Based on these forces the United Nations Environmental Programme (UNEP) has categorized natural disasters into three categories:

- **Meteorological disasters:** Drought, cyclone, cold and heat wave etc.
- **Topological disasters:** Earthquake, avalanches, landslides etc.
- **Biological disasters:** Insect, swarms, virus and epidemics of communicative diseases.

Disasters include

- Tornadoes
- Fires
- Hurricanes
- Floods/Sea Surges /Tsunamis
- Snowstorms
- Earthquakes
- Landslides
- Pollution
- Heatwaves, Cold wave

- Drought
- Building collapse
- Nuclear accidents
- Explosions
- War
- Water contamination and Existing or anticipated food shortages.

1.4 Types of Disaster

Disasters are broadly classified into two types – *Natural* and *Manmade.*

According to World Disaster Report (2003), natural disasters include hydro-meteorological disasters and geophysical disasters. The hydro-meteorological disasters include landslides/avalanches; droughts/famines; extreme temperatures and heat waves; floods; hurricanes; forest/scrub fires; windstorms; and others (insect infestation and waves/surges). The geophysical disasters include earthquakes and volcanic eruptions.

Based on the devastation, these are further classified into major/minor natural disaster and major/minor manmade disasters. Some of the disasters are listed below:

Major natural disasters:

- Flood
- Cyclone
- Drought
- Earthquake

Minor natural disasters:

- Cold wave
- Thunderstorms
- Heatwaves
- Frost
- Cloud burst

Major man-made disasters:

- Setting of fires
- Deforestation
- Pollution due to many human activities
- Industrial/Air pollution.
- Wars

Minor man-made disasters:

- Road /train/air accidents, riots
- Food poisoning
- Industrial disaster/crisis
- Biological disaster
- Soil and water pollution.

1.5 Effect of Disaster

Disaster is causing widespread devastation to life and property in a region in a few seconds to minutes. The damage caused by the disaster is immeasurable and varies with the location, type of settlement, type of the earth surface and financial capacity of the affected people. According to India's tenth five-year plan, natural disasters have affected nearly 6 per cent of the population and 24 per cent of deaths caused by disasters. Between 1996 and 2001, 2 per cent of national gross domestic production (GDP) was lost because of natural disasters, and nearly 12 per cent of Government revenue was spent on relief, rehabilitation and reconstruction during the same period. According to World Bank report, natural disasters pose a major hurdle on the path of economic development in India. The expenditure for management of disaster was very low (<100 US$) during 1965 – 1974, and increased after 1974, and increased during the years 1974–75 (900US$) and 1989–90 (nearly 5000US$) (Figure 1.6).

Reported catastrophe losses in India, 1965–2001

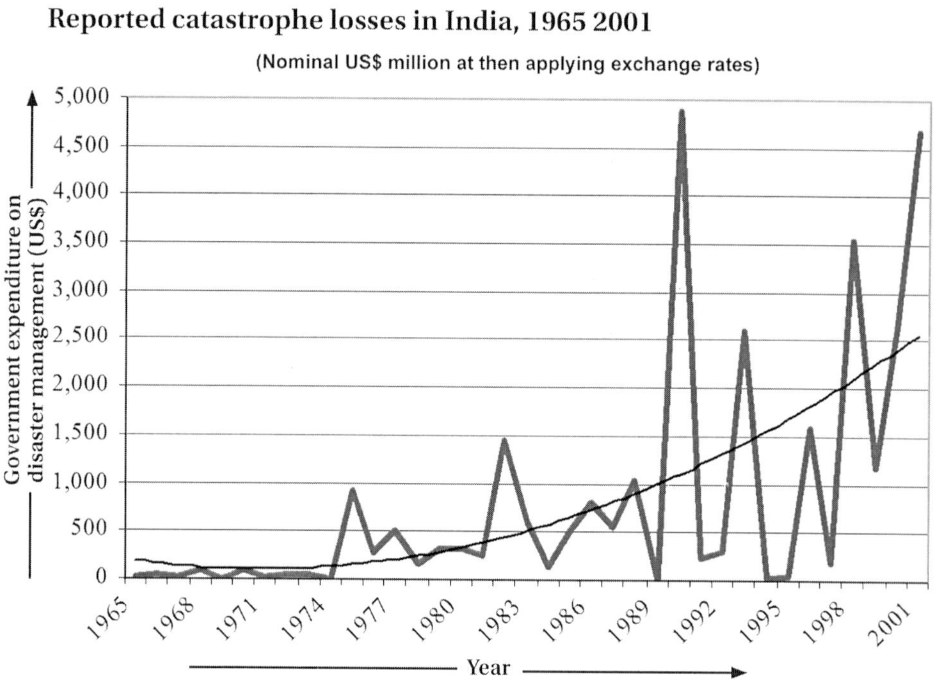

Figure 1.6 : Government revenue spent on disasters management [*Source:* MoAC]

1.6 Hazard

A Hazard means the potential to damage, and it is the event that may lead to any disaster. When it occurs, it damages the property and life, and becomes a disaster. For example, a flood is an event means a hazard. When it occurs, it damages homes and washes away peoples, and affects their livelihood. Then, the flood becomes a disaster. But if; people are evacuated along with valuables to a safe shelter, and houses are built with adequate features to make them flood-resistant, the flood remains a hazard and does not become a disaster.

Hazard may be defined as "a dangerous condition or event, that threat or have the potential for causing injury to life or damage to property or the environment." The word 'hazard' owes its origin to the word 'hasard' in old French and '*az-zahr*' in Arabic meaning 'chance' or 'luck'.

Natural Hazard: A physical event or phenomena which may cause injury or loss of life, damage to property, social and economic disruption or environmental degradation.

1.6.1 Hazard Classification

There are several methods of classifying a hazard, and depends on the variation on the factors of "likelihood" of the hazard turning into an incident and the "seriousness" of the incident.

A common method is to score both likelihood and seriousness on a numerical scale (with the most likely and most serious scoring highest) and multiplying one by the other to reach a comparative score.

Hazards are sometimes classified into the following modes:

Dormant – It is the situation and has the potential to be hazardous, but no people, property, or environment is currently affected by this situation. For instance, a hillside may be unstable, and has the potential for a landslide, but there is nothing below or on the hillside that could be affected.

Active – A harmful incident involving the hazard has actually occurred. Like an accident, emergency, incident, or disaster.

A natural phenomenon that occurs in a populated area is known as **hazardous event**. Hazard has also accompanied by armed and they are people, property, or environment.

Each hazard has its own characteristics. Every hazard has its nature, causes and effects, and each and every hazards vary from one another. Hazards can be grouped into two broad categories namely **natural** and **manmade** (Table 1.2).

Table 1. 2 : Frequently occurring natural hazards in India

Type	Location /Area	Affected population (in million)
Floods	8 major river valleys over 40 million hectares of area in the entire country	260
Drought	Spared in 14 stars of Andhra Prdesh, Bihar, Gujarat, Haryana, Jammu and Kasmir, Karnatka, Madhya Pradesh, Mahasrastra, Odisha, Rajasthan, Tamil Nadu, Uttar Pradesh, West Bengal and Himachal Prdesh covering a total of 116 districts and 740 blocks.	86
Earthquake	Nearly 55 percent of the total area of the country falling in the seismic zone IV and V	400
Cyclones	Entire 5700 km long coastline of Southern, Peninsula of India covering 9 states *viz.*, Gujrat, Maharastra Goa, Karnatka, Kerala, Tamil Nadu, Andhra Pradesh, Orisha and West Bengal.	10
Landslide	Entire sub-Himalayan region and Western Ghats	10

1.6.2 Types of Hazard

Hazards are generally of four types, physical hazards, chemical hazards, biological hazards and allergenic hazard.

Atmospheric :

- Cloud burst
- Squall
- Cyclones
- Lightning
- Tropical storms/low pressure area

Seismic :

- Fault ruptures
- Ground shaking
- Lateral spreading
- Liquefaction

Tsunamis Hydrologic :

- Coastal flooding
- Desertification
- Salinization
- Drought
- Erosion and sedimentation
- River flooding
- Storm surges

Other Geologic/Hydrologic :

- Debris avalanches
- Landslides
- Rock falls
- Submarine slides
- Subsidence

Volcanic :

- Ash domes
- Gases
- Lava flows
- Mudflows
- Projectiles and lateral blasts
- Pyroclastic flows

Wildfire :

- Brush
- Forest
- Grass
- Savannah

Hazard is the potential to cause harm.

Sometimes a hazard is referred to the actual harm it caused rather than its effects. For example, TB-causing bacteria would be considered the "hazard", but some time the effect of the disease tuberculosis (TB) might be called a hazard.

A common way to classify hazards is by category *i.e.* Biological, Chemical, Ergonomic, Physical, Psychosocial and safty, and denoted by BCEPS:

- **Biological** — bacteria, viruses, insects, plants, birds, animals, and humans, etc.,
- **Chemical** — depends chemical and toxic properties of the chemical.
- **Ergonomic** — repetitive movements, improper set up of workstation, etc.,
- **Physical** — radiation, magnetic fields, pressure extremes (high pressure or vacuum), noise, etc.,
- **Psychosocial** — stress, violence, etc.,
- **Safety** — slipping/tripping hazards, inappropriate machine guarding, equipment malfunctions or breakdowns.

Hazard is a destructive phenomenon or event and it could be earthquakes, tropical storms, volcanic eruption or industrial events and these events themselves are the hazards, are not disasters. When a hazards affect populations directly or indirectly, through the destruction of belongings, lives, crops, water sources or physical infrastructure than they become disasters.

Hazard management is like a problem-solving process and involves to define the problems (identifying hazards), to gather information about them (risk assessment) and to solve them (risk control). This is followed by to check (evaluation) and to review the whole process (review) after a period of time or when there is a change.

Many a time, natural and manmade hazards are combined and become a disaster. For example, flooding may be caused either by heavy rains, landslide or blocking of drains with human waste. The degree of damage by a hazard also depends on the social-economical conditions and its exposure.

Exposure = duration and/or extent of a hazard

An earthquake is disastrous only when it affects people, their properties and activities. Thus, disaster occurs only when hazards and vulnerability meet., but it is also to be noted that with greater capacity of the individual/community and environment to face these disasters, the impact of a hazard may be reduced. Therefore, we need to understand the three major components namely hazard, vulnerability and its capacity with suitable examples to have a basic understanding of disaster management.

Hazards are broadly be classified as below:

1. **Natural hazards:** are caused by natural phenomena and have a natural origin (like meteorological, geological or even biological origin). Examples of natural hazards are cyclones, tsunamis, earthquake and volcanic eruption.
2. **Manmade hazards:** are caused by human activities including negligence. Manmade hazards are associated with industries or energy generation facilities and include explosions, leakage of toxic waste, pollution, dam failure, wars or civil strife etc. *e.g.* Bhopal gas leakage in December 2002.

1.6.3 The Nature of the Hazard

Rapid Onset

The speed of onset of a hazard is an important variable, which involves its warning time. Like extreme earthquakes, landslides, and flash floods give virtually no warning. Less extreme are tsunamis, which typically have warning periods of minutes or hours as well as hurricanes and floods, where the likelihood of occurrence is known for several hours or days in advance. When the warning time of a hazard is very less in hours as well as day than this type of hazards are known as rapid onset hazard like cloud burst.

Slow Onset

Other hazards, such as drought, desertification, and subsidence are slow occurrence over a period of months or years. Hazards such as erosion/ sedimentation have varying lead times: Earthquake develop over many years. Hazards; which involve a wide scale of warning time are known as slow onset hazard. Like drought, erosion, deforestation, pollution etc.

Mention the true and false against each of the questions given below.

a) An Indo-Gangetic plain is drought prone region of India.

b) Flash flood is rapid onset of disaster.

c) Poor are mostly vulnerable to any type of disaster in India.

d) Hazards are not having potential to cause harm.

e) The deadly cyclone which was occurred during 1977 in Andhra Pradesh killed 200 people.

1.7 Risk

Risk is the measurement of the expected losses caused by any hazards in an area in a specific time period. Risk depends on two factors; the probability of occurrence of hazards and the measurement of losses by a hazard in an area.

Risk is the likelihood of harm (in defined circumstances, and usually qualified by some statement of the severity of the harm). Risk is the chance or probability that a person will be harmed or experience an adverse health effect if exposed to a hazard.

Risk is a function of the probability of particular occurrences and the losses which would cause. The level of risk depends on:

- Nature of the Hazard
- Vulnerability of the elements which are affected
- Economic value of those elements.

> ***Risk = Hazard × Vulnerability(-) Capacity***
> *This score can then be used to identify which hazards may need to be mitigated. A low score on the likelihood of occurrence may mean that the hazard is dormant, whereas a high score would indicate that it may be an "active" hazard.*

Hazard and vulnerability interact together to create risk.

Risk is the likelihood of harm. Risk is the chance or probability that a person will be harmed or experience an adverse effect if exposed to a hazard. Factors influencing the degree of risk are:

- how much a person is exposed to a hazardous thing or condition,
- how the person is exposed, and
- how severe are the effects under the conditions of exposure

A community/locality is said to be at 'risk' when it is exposed to hazards and is likely to be adversely affected by its impact.

1.7.1 Risk Factors

The risk factors are either inherited (genetically) or acquired, but their occurrence is not defined, which one comes first. For example, the persons who may have suffered for eczema, dermatitis, and asthma in childhood may be considered as having a high-risk factor, and the likelihood of developing asthma from exposure to say, animals, later on in life. Smokers may also be at a higher risk of developing cancer, or some forms of occupational asthma.

1.7.2 Risk Assessment

Risk assessment is a process and involves:

- Identification of hazards,
- Analyze or evaluate the risk associated with that hazard, and
- Determine appropriate ways to eliminate or control the hazard.

1.7.3 Risk Characterization

The characterization of risk has both quantitative and qualitative components.

It is clear that the degree of exposure is a very important determinant of risk, *e.g.*, a low exposure to something that is highly hazardous may result in a low risk. High exposure to something of incredibly low hazard may result in a moderate or even high risk, therefore every reasonable attempt is made to quantify and exposure in order to attribute a measure at risk to it.

1.8 Vulnerability

It is defined as "the extent to which a community, structure, service, or geographic area is likely to be damaged on account of their nature, construction and proximity to hazardous terrain or a disaster-prone area or disrupted by the impact of a particular hazard."

Vulnerability, in simple terms is the potential for loss to an individual, community or place because of a disaster, which is affected by geographical as well as social conditions

Vulnerability = susceptibility to damage or harm by a hazard

Vulnerability may be affected by many factors like forces in society, geographic location, social and economic factors and so on so forth. The extent of vulnerability depends very much not only on characteristics of the person or household, but on the nature of the hazard. Vulnerability of persons living in a substandard apartment house would differ from their vulnerability to an earthquake to a earthquake-resistant apartment/buildings. Vulnerabilities to the same hazard will differ between poor and the rich.

1.8.1 Socio-economic Vulnerability

The degree or intensity of a hazard to which a population will be affected depends on the physical components of vulnerability and also on the socioeconomic conditions. The socio-economic condition of the people also determines the intensity of the impact. For example, poor people who are living in the seacoast area/region are generally at risk and lose their shelters, whenever there is strong wind or cyclone. Due to their poverty, they too are not able to rebuild their houses. Hence risk is more for resource - poor people than the resource rich people. Poverty is perhaps the major vulnerability of a population.

Social vulnerability

- *90 percent poor, marginal and small farmers are affected by disasters.*
- *Pockets of disasters are semi-arid, arid regions, flood plains of indo-gangetic and other river basins, mountainous regions of NE and Himalayan region.*

Economic impacts

- *4.5 to 10 percent loss in national GDP.*
- *Loss of revenue to the extent of 10–20 percent of state GDP.*
- *15 to 50 percent reduction in food grain production.*

1.9 Defining the Scope of Disaster

There is no single measure that can capture the full scope of a disaster. Every person has their scope and the individual will consider the impact on his or her family and livelihood.

Disaster managers will assess the speed and success of the disaster response. Economists will measure physical loss to houses and buildings and loss of production. Health workers will consider the resources required to contain an outbreak of diseases. Others may focus on the nature of the hazard, the social consequences and the impact on specific elements of the infrastructure. A common measure is the number of people killed or affected.

1.10 The Disaster Cycle

Disasters also have many phases and each phase may be followed by sequential manner to another phase and hence thought to be a cycle.

It may occur with or without a *warning phase*. Some of the common phases are:

i) **Preparedness phase:** Before the disaster all measures to reduce the impact of a disaster.

ii) **Emergency phase:** Immediately after the disaster, an emergency services are required.

iii) **Response phase:** Just after the disaster immediate and early actions are required.

iv) **Reconstruction (rehabilitation) phase:** After the disaster to provide basic facilities to the people.

v) **Mitigating:** To prepare or to prevent the disasters in future.

1.11 Difference between an Emergency and a Disaster situation

An emergency and a disaster are two different situations:

- An ***emergency*** is a situation in which the community is capable of coping. It is a situation generated by the real or imminent occurrence of an event that requires immediate attention and help with nearby available resources.
- A ***disaster*** is a situation in which the community is incapable of coping. A event which causes intense negative impacts on people, goods, services or the environment, exceeding the affected community's capability to respond; therefore the community seeks the assistance of government and international agencies.

1.12 Development and Disaster

The link between disasters and development has been realized. It is two-way link. Disasters have an impact on development and developments also contribute to the disasters. Development affects economic status, and all disasters have negative economic effects, but the developing nation with a fragile economy is more vulnerable. Disasters have linkages with climate change and natural resources use (Figure 1.7).

> *Disaster management is basically 'disaster risk management'. Disaster risk management includes all measures which reduce disaster-related losses of life, property or assets either by reducing the hazard or by the vulnerability of the elements at risk.*

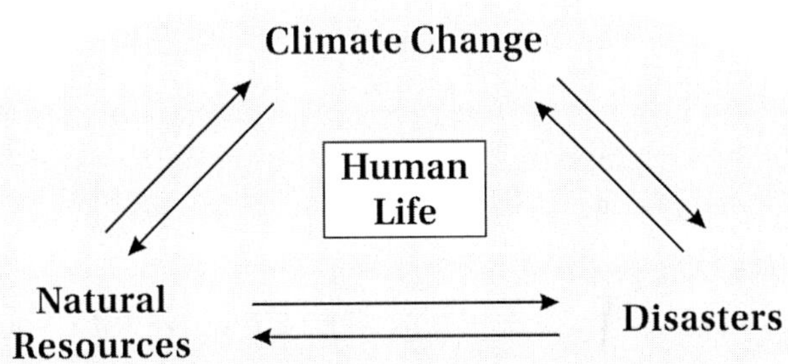

Figure 1.7 : Linkages of disasters

Questions for Chapter evaluation

Descriptive questions

1. Define Disaster.
2. Define hazard.
3. What is the difference between a disaster and a hazard?
4. Define Risk.
5. Name the categories of disaster.
6. List the types of natural disaster.
7. What are the examples of rapid-onset disaster?
8. What are the examples of slow-onset natural disaster?
9. What do you mean by the vulnerability?
10. Is the social and economical vulnerability associated with disaster? Explain with examples.
11. What is disaster cycle?
12. List out the major disasters events of Indian subcontinent.
13. What is difference between disaster and emergency?
14. How risk is different from vulnerability?
15. List the vulnerable pockets of your area.

Objective questions

Fill in the blanks

i) According to the United Nations between 1992 and 2000 there were approximate .... disasters per year throughout the world.

ii) A disaster disturbs the ....

iii) Out of the 29 states and 7 union territories in India, the numbers of disaster-prone states are ....

iv) .... percent of landmass is prone to earthquakes in India

v) .... percent of landmass is prone to flood in India. The disaster cycle has how many phases

Mention true and false against each of the questions given below.

a) An Indo-Gangetic plain is drought-prone region of India.

b) Flash flood is rapid onset of disaster.

c) Poor are most vulnerable to any type of disaster in India.

d) Hazards are not having potential to cause harm.

e) The deadly cyclone, which was occurred during 1977 in Andhra Pradesh killed 200 people.

CHAPTER 2

NATURAL DISASTER

A natural disaster (ND) is a result of the natural event with catastrophic consequences for living things in the affected area and its surroundings. Earthquakes, volcanoes, floods, drought, cyclones, heat and cold wave and landslides/avalanches are natural events having disastrous impacts, when they affect human settlements and life. Human societies have witnessed a large number of such natural hazards in different parts of the world as well as India and have tried to learn the lesson to minimize the loss to some extent. Drought, floods, cyclones, heat and cold wave affected region of India is shown in Figure 2.1.

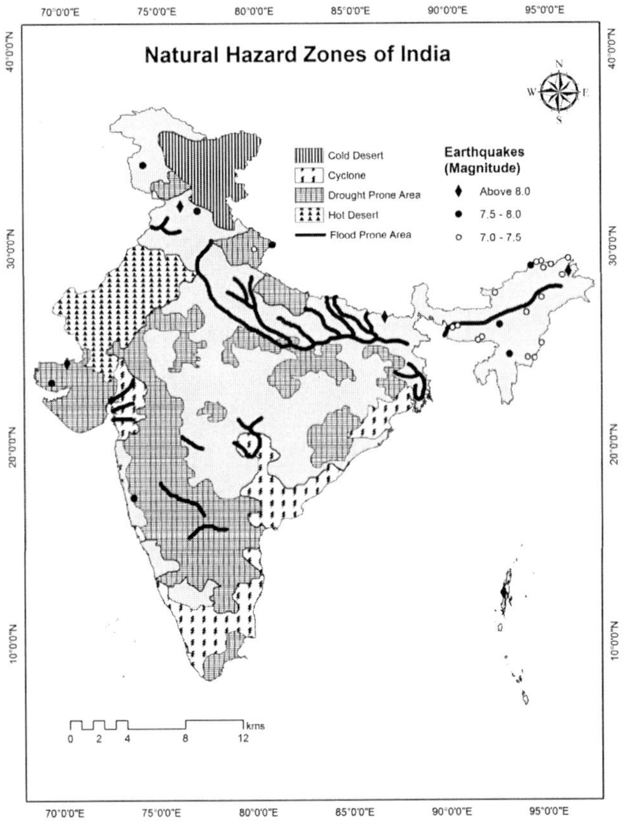

Figure 2.1 : Natural hazard zone of India

2.1 Severity of Natural Disasters

Natural disasters occur in all shapes and size ranging from a small scale flood to a large scale tsunami. Currently, existing scales for natural disaster define severity levels in terms of intensity. Overall, intensity levels are not the best way to describe the severity levels of disasters because they are an indication only of strength, but not the impact of the disaster. Looking at the risk associated with natural disaster, their severity scale based on its intensity was made on a 0–10 scale and presented in Figure 2.2. The highest severity is of an earthquake (9.9), while the smallest intensity is of Tornado 6.7.

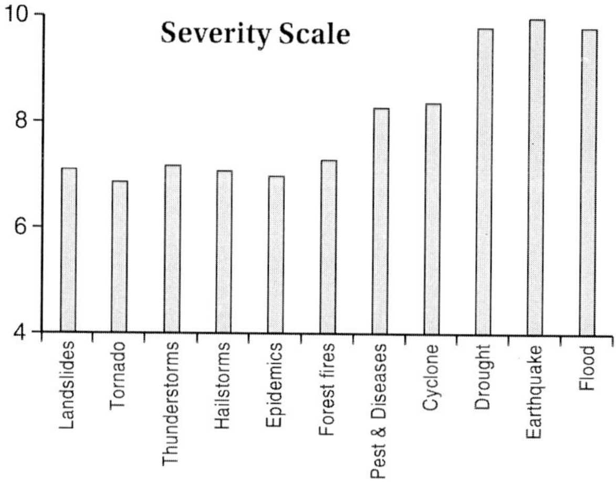

Figure 2.2 : Severity scale of natural disasters [*Source:* United Nations world conference on disaster risk reduction (WCDRR) in 2005 in Japan]

The impact depends on where a disaster occurs, i.e., a populated city or rural area. In addition, there are different types of scales for different disasters. Earthquakes are measured using the Richter scale, Volcanic eruptions using the Volcanic Eruption Intensity (VEI) and Tornadoes using Enhanced Fujita Scale (EF-scale).

Disaster scope has been recently presented by M. had-el-Hal (2008) and shown in table 2.1. Disaster scope has five levels according to its impacts on life and area.

Table 2.1: Disaster scope and its affected area

Scope	Types	Number of casualties	Geographical area affected (km^2)
I	Small disaster	<10	<1
II	Medium disaster	10–100	1–10
III	Large disaster	100–1000	10–100
IV	Enormous disaster	1000–10000	100–1000
V	Gargantuan disaster	>10000	>1000

[*Source : M. Gad-el-Hak ed. (2008) The Art and Science of large scale disasters*].

The Indian subcontinent has witnessed of many major disasters like Uttarkashi earthquake (1991), Killari earthquake (1993), Koyana earthquake (1997), Chamoli earthquake (1999), and Bhuj earthquake (2001), Jammu and Kashmir earthquake (2005). Some of the major disasters, which were occurred in India are shown in Table 2.2.

Table 2.2 : List of Major Natural Disasters in the known history of India

S.N.	Name of Event	Year	Fatalities
1.	Bengal Earthquake	1737	300,000
2.	Bengal Cyclone	1864	60,000
3.	The Great Famine of Southern India	1876–1878	5.5 million
4.	Maharashtra Cyclone	1882	100,000
5.	The Great Indian famine	1896–1897	1.25 million to 10 million
6.	Kangra Earthquake	1905	20,000
7.	Bihar Earthquake	1934	6,000
8.	Bengal Cyclone	1970	500,000 (include Pakistan and Bangladesh also)
9.	Drought	1972	200 million people affected
10.	Andhra Pradesh Cyclone	1977	10,000
11.	Latur Earthquake	1993	7,928 death and 30,000 injured
12.	Odisha Super Cyclone	1999	10,000
13.	Gujarat Earthquake	2001	25,000
14.	Indian Ocean Tsunami	2004	10,749 deaths 5,640 persons missing
15.	Kashmir Earthquake	2005	86000 deaths (include Kashmir and Pakistan)
16.	Kosi Floods	2008	527
17.	Cyclone Nisha of Tamil Nadu	2008	204

2.2 Earthquakes

Earthquake is one of the most destructive natural hazard. It is caused by the sudden shaking of the earth crust. The impact of an earthquake is sudden and there is hardly any warning, making it impossible to predict. It may occur at any time in a year, in a month, in a week, or day, hours, minutes and seconds. An earthquake is resultant of a sudden movement of the earth's crust or shaking of the earth. Shaking of the earth either caused by the moment in earth's lithosphere, the release of energy within rocks or movement along geological faults, or by the movement of magma in volcanic areas. They can destroy buildings and infrastructure in seconds, killing or injuring the inhabitants. Earthquakes not only destroy the entire habitation, but may de-stabilize the government, economy and social structure of the country. Earthquakes represent a severe threat due to the irregular occurrence, lack of adequate forecasting, and risk and severity associated with the hazard.

2.2.1 The Composition and Structure of Earth

The earth-based on composition has been divided into three main parts into the crust, mantle, and core. The crust makes up less than 1 percent of Earth by mass, consisting of oceanic crust and continental crust is often more felsic rock. The mantle is hot and represents about 68 percent of Earth's mass. Finally, the core is mostly iron metal. The core makes up about 31 percent of the Earth. Each of these layers can be further divided into two parts: the inner and outer core, the upper and lower mantle and the continental and oceanic crust. Both the inner and outer core is made up of mostly iron and a little bit of nickel. It has a mean diameter of 12742 km with a radius of 6371 km (Figure 2.3).

Lithosphere and asthenosphere are divisions based on mechanical properties. The lithosphere is composed of both the crust and the portion of the upper mantle that behaves as a brittle, rigid solid. It is the outermost mechanical layer, which behaves as a brittle, rigid solid. The lithosphere is about 100 kilometers thick. The asthenosphere is partially molten upper mantle material that behaves plastically and can flow.

i) Crust

The crust is the outermost layer of the Earth. This is the outside layer of the earth and is made of solid rock, mostly basalt and granite. It is mostly made up of the lighter elements, silicon, oxygen, aluminium. Due to this, it is known as sial (Silicon = Si; Aluminium = Al) or felsic. There are two different types of crust: thin oceanic crust (5–10 Km) that underlies the ocean basins, and thicker continental crust (30–50 Km) that underlies the continents. These two different types of crust are made up of different types of rock. The thin oceanic crust is composed of primarily of basalt, and the thicker continental crust is composed primarily of granite. The low density of the thick continental crust allows it to "float" in high relief on the much higher density mantle below. The thickness of the crust varies under the oceanic and continental areas. The crust made up of heavier rocks having a density of 3 g/cm^3. The mean density of material in the oceanic crust is 2.7 g/cm^3.

ii) Mantle

The mantle is the layer of the Earth right below the crust. It is made up mostly of oxygen, silicon and the heavier element magnesium. It is known as sima (Si for Silicon + Ma for Magnesium) or mafic. The two most important things about the mantle are: (1) it is made of solid rock, and (2) it is hot. The mantle itself is divided into layers. The uppermost part of the mantle is solid and forms the base of the crust. It is made of the heavy rock peridotite. It has a density higher than the crust portion. The thickness ranges from 10–200 km. The mantle extends from Moho's discontinuity to a depth of 2,900 km. The asthenosphere is the upper portion of Mantle. It is the chief source of magma that finds its way to the surface during volcanic eruptions. The continental and oceanic plates include both the crust proper and this uppermost solid layer of the mantle. Together this mass makes up the lithosphere. The lithosphere plates float on the semi-liquid aesthenosphere below.

1. Upper aesthenosphere: magma
2. Lower aesthenosphere
3. Lower mantle

iii) Core

At the earth's centre lies a dense metallic core. The density of Earth's surface layers is much less than the overall density of the planet, as calculated from the planet's rotation. If the surface layers are less dense than average, then the interior must be denser than average. Calculations indicate that the core is about 85 percent iron metal with nickel metal making up much of the remaining 15 percent. The core-mantle boundary is positioned at the depth of 2,900 km. The inner core is in the solid-state, whereas the outer core is in the liquid state. The core is made up of very heavy material mostly constituted by nickel and iron. Hence it is also called the "nife" layer. The earth's core is made up of solid iron and nickel, and is at about 5000–6000°C. Outer core is a liquid layer below the mantle. Inner core is the very centre of the Earth.

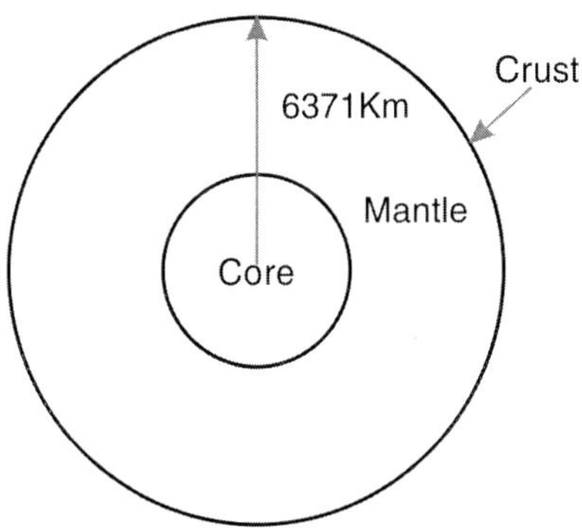

Figure 2.3 : Earth inner view of core, Mantle and crust

2.2.2 Causes of Earthquake

The outer shell of the earth, the lithosphere is broken up into tectonic plates. The Earth is made of large plates of land that float over an ocean of semi molten magma. If any plate with an area > 20 million km^2 is known as a major plate. The seven major plates are shown in table 2.3. When, two plates collide with each other huge amount of energy released and thus shake the earth crust, an earthquake is observed. There are also other events due to which there may be movement (1–4cm/year to 5–10cm/year) in plates on earth crust.

Table 2.3: The earth plates and associated area

Name of the plate	Area in Km^2
Pacific	103,300,000
North American	75,900,000
Eurasian	67,800,000
African	61,300,000
Antarctic	60,900,000
Indo-Australian	58,900,000
Australian	47,000,000
Indian	11,900,000
South American	43,600,000

There are numerous causes of earthquakes and listed below:

- Plate movement over mantle of the earth
- Rupturing along faults, or sliding of plates over one another
- Release of magma on earth crust
- Sliding of one plate over other plates
- Landslide in or over ocean floor.

Earthquakes also caused, when underground rock suddenly breaks/slip along a fault. Volcanic eruption or liquefaction of material creates strong stress along faults (Figure 2.4). Due to the release of energy in form waves (seismic) shaken the earth crust. The tectonic plates are always moving on earth mantle and some time struck or touched each other and makes stress on edge to overcome the friction and thus releases energy in form of wave travel through earth crust and caused shaking of earth.

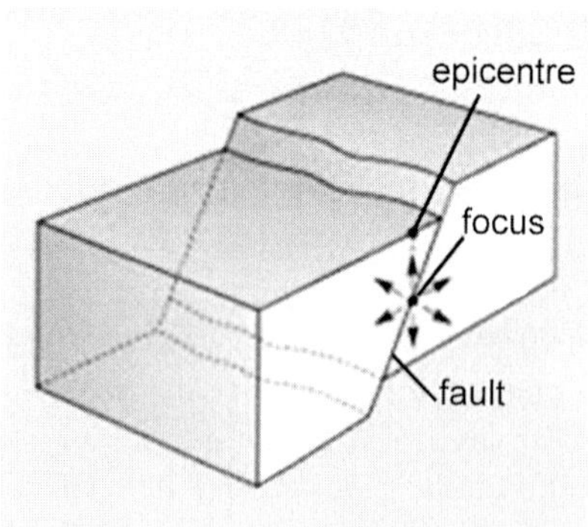

Figure 2.4 : Sliding of rocks along fault and epicentre

Earthquakes are more likely to occur along faults. Earthquakes and volcanoes occur most commonly at the collision zone between two moving tectonic plates across each other.

The earth's crust is a rocky layer of varying thickness ranging from a depth of about 10 kilometers under the sea to 65 kilometers under the continents. The crust is not one piece, but consists of portions called *'plates'* (Figure 2.5). There are seven major plates and several minor plates which are moving a few inches a year and riding on semi-molten layers of rock underneath the crust. When these plates contact each other, stress arises near the area of the contact zone. These stresses can be due to the type of movement along the plate's boundaries:

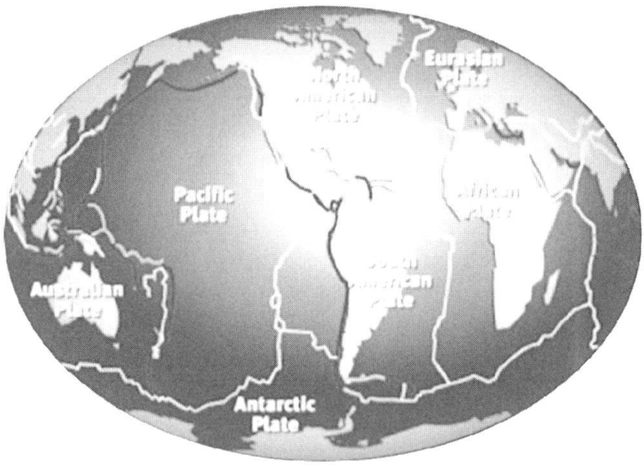

Figure 2.5 : Earth plates on earth crust
[*Source:* Online search free download image]

a) Pulling away from each other,

b) Pushing against one another and

c) Sliding relative to each other.

All these movements may cause earthquakes. The volcanic eruptions, rock falls, landslides and explosions can also be caused shaking in the earth crust.

The areas of stress at plate boundaries, which release accumulated energy by slipping or rupturing are known as 'faults'. The fault rupture generates vibration called seismic waves, which radiates from the focus in all directions. The point of rupture is called the 'focus' and may be located near the surface or deep below it. The point on the surface directly above the focus is termed as the 'epicenter' of the earthquake.

2.2.3 General Characteristics

During earthquake, vibrations create in a variety of frequencies and velocities. The actual rupture process may last for a few seconds to as long as one minute for a major earthquake. The ground-shaking is caused by 'body waves' and 'surface wave'.

Earthquakes can be of three types based on the focal depth:

Deep earthquake : 300 to 700 kms from the earth surface

Medium earthquake : 60 to 300 kms

Shallow earthquake : less than 60 kms

The deep focus earthquakes are rarely destructive because by the time the waves reach the surface its impact is reduced very much. Shallow focus earthquakes are more common and are extremely damaging because of their proximity to the surface. Shallower earthquakes occur frequently than deep earthquake.

2.2.4 Measuring of Earthquakes

Earthquakes are measured on the two distinctively different scales of measurement, *i.e.,* either by measuring their magnitude or their intensity. Thus there are two scales used for measurement of earthquake, *i.e.,* magnitude scale and intensity scale.

i) Magnitude Scale or Richter scale

Earthquake magnitude or amount of energy released is determined by an instrument, which continuously records ground vibration is known as a seismograph. The scale on which magnitude is measured was developed by a seismologist named Charles Richter and popularly known as Richter Scale for magnitude measurement. This scale is levelled from 1 to 10 and each level is 10 times more than the previous level (Table 2.4). An earthquake of magnitude 3 is the smallest normally felt by humans, but above 6 magnitude scale, earthquakes are very destructive. The largest earthquake that has been recorded in the past was 9.25 (Alaska, 1969 and Chile, 1960).

Table 2.4: Richter scale for earthquake release energy measurement

Level of scale	Earthquake type	Occurrence frequency	Observed Cycle
1	Small	Daily	Every minute
2	Small	Daily	Every hour
3	Small	Daily	Every day
4	Small	Daily	Every Week
5	Moderate	Monthly	Every 10 years
6	Moderate	Monthly	Every 30 years
7	Major	Monthly	Every 50 years
8	Great	Yearly	Every 100 years
9	Great	Yearly	Every 300 years
10	Super	Rarely	Every 1000 years

ii) Intensity scale or Mercalli Scale

The second type of scale is known as intensity scale, which measures the effect of an earthquake on the objects where it occurs means in this scale indirect measurement is done, without the use of any instrument. This is most widely used scale and was developed by Mercalli an Italian seismologist in 1902. The scale was extended and modified to suit the modern times and known as the Modified

Mercalli Scale, which expresses the intensity of earthquake effect on people, structure and the earth's surface and ranged from I to XII (Table 2.5). With an intensity of VI and below this, most of the people can feel the shake, but with an intensity of XII, there is high impact on structures and buildings, which are collapsing totally and make total disruption in normal life.

Table 2.5 : Modified Mercalli scale for intensity measurement

Level of scale	Observed effects
I	Not felt
II	Felt only by persons at rest or doing nothing or sleeping
III–IV	Felt by persons who are inside a building or house; no damage observed
V-VI	Generally felt by all; damage to plaster, cracks in uncemented walls
VII	People run from building/homes; damage to poorly built structures
VIII	Well-built structures slightly damaged; poorly built structures suffer major damage
IX	Building collapsed completely and shifted off foundations
X	Some well-built structures destroyed
XI	Few masonry structures remain standing, bridges destroyed
XII	Damage total, waves were seen on ground; objects thrown into air

2.2.5 Adverse Effects

The effect of an earthquake is very wide and diverse, which damage property, buildings, life, surface features and thus affect transportation, communication and supply of commodities. It creates a huge loss to food, shelter and health system and water supply in the affected areas were observed. Secondary effects include fires, dam failure and landslides and flooding.

Physical damage and death : Physical damage are those damages, which occur to human settlement, buildings, structures and infrastructure, especially bridges, elevated roads, railways (Figure 2.6), water towers, pipelines, electrical generating facilities and much greater damage to already weakened structures.

Some examples of earthquakes damage are given below :

Latur earthquake, which was occurred during 1993, caused 8000 deaths and 2 lakhs houses are either damaged or destroyed. Bhuj earthquake, which occurred during 2001, caused 13,800 deaths and 1,000,000 houses are either damaged or destroyed. Jammu and Kashmir earthquake, which was occurred during 2005 caused 1300 deaths and damaged 41 thousand houses.

Figure 2.6 : Physical damage of building and dam

2.2.6 Earthquakes Risk Zone (ERZ) in India

In our country, some area/zones are very risky and frequent earthquakes are recorded in this zone and they are known as earthquake risk zone of India. The 12 per cent area is prone to very severe earthquakes,18 per cent area to severe earthquakes and 25 per cent area to damageable earthquakes, thus a total of 55 per cent area of the country comes under earthquake zones. Based on the frequency of occurrence of earthquakes, Bureau of Indian Standards has grouped the country into four seismic zones namely Zone II, III, IV and V and shown in Figure 2.7. The zone I is an area, which is free from the occurrence of earthquakes. Zone IV and zone V are rated as the most seismically prone region, while zone II is the least. The Modified Mercalli intensity (MMI), which measures the impact of the earthquakes on the surface of the earth, broadly associated with various zones, is shown in table 2.6. Most of the region of India lies in zone III. Madhya Pradesh and Chhattisgarh state lie in zone II. The Delhi lie in zone IV, whereas other the mega cities like Mumbai and Chennai are in zone III.

The biggest earthquakes occur in the Andaman and Nicobar Islands, Kutch, Himachal and the North-East. The Himalayan regions are particularly prone to earthquakes. The seven sister states of North East - Assam, Arunachal Pradesh, Nagaland, Manipur, Mizoram, Tripura and Meghalaya; Andaman and Nicobar Islands; and parts of 6 other states in the North/North-West (Jammu and Kashmir, Uttaranchal, Bihar) and West (Gujarat), are in Seismic Zone V. Lists of major earthquakes in India and its impacts is shown in Table 2.7.

Table 2.6 : Seismic zone and affected area of India

Seismic Zone	Affected area of percent of total area
II (Low intensity zone)	43%
III (Moderate intensity zone)	27%
IV (Severe intensity zone)	18%
V (Very severe intensity zone)	12%

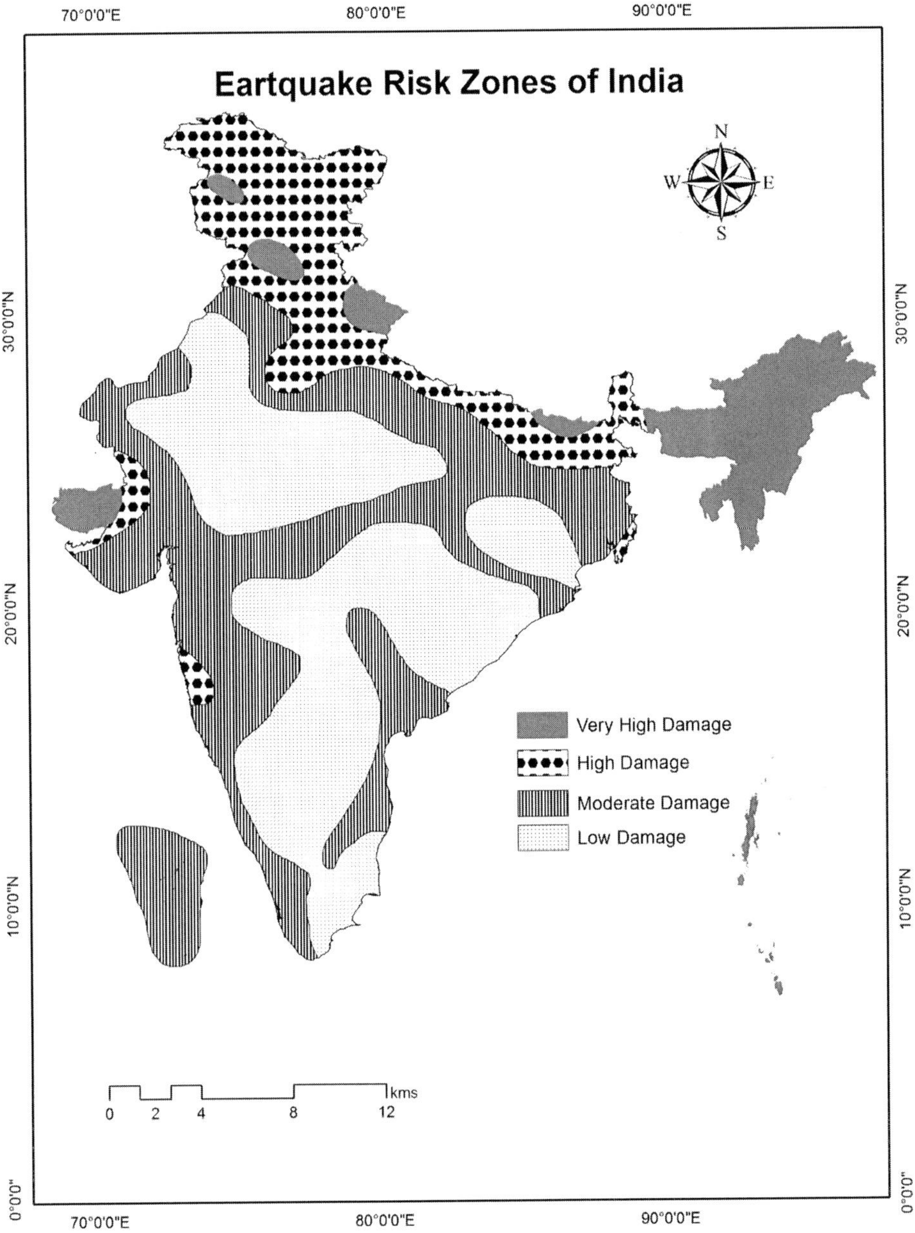

Figure 2.7 : Earthquake risk zones of India

The Indian subcontinent has a history of devastating earthquakes and some of them are given in table 2.8. According to plate tectonic theory each major plates are moving and when they meet each other at their edge, put a strong pressure and after that released their accumulated energy into form of a wave. The Indian plate is driving towards Asian plate at a rate of approximately 47 mm/year and this is the major reason for high frequency and intensity of the earthquakes.

Table 2.7: Major earthquakes of India and their damage

Place	Date	Magnitude	Affected area	Death
Bhuj, Gujarat	26th January, 2001	7.7	Bhuj, Anjaar, Vondh and Bhachau	20,000
Bihar	15th January, 1934	8.1	Purnea, Munger, Muzaffarpur and Champaran	30,000
Maharashtra	30th Sept., 1993	6.4	Latur	20,000
Medog, Assam	15th August, 1950	8.6	Assam and Tibet	1,500
Uttarkashi	20th October, 1991	6.1	Uttarkashi, Chamoli and Tehri	1,000

Table 2.8 : List of major earthquakes in India after independence

Year	Location	Magnitude of 6 and above
1950	Arunachal Pradesh - China Border	8.5
1956	Anjar, Gujarat	7.0
1967	Koyna, Maharashtra	6.5
1975	Kinnaur, Himachal Pradesh	6.2
1988	Manipur - Myanmar Boarder	6.6
1988	Bihar - Nepal Border	6.4
1991	Uttarkashi - Uttar Pradesh Hills	6.0
1993	Latur - Maharashtra	6.3
1997	Jabalpur, Madhya Pradesh	6.0
1999	Chamoli, Uttar Pradesh	6.8
2001	Bhuj, Gujarat	6.9
2005	Muzaffarabad (Pakistan) Impact in Jammu & Kashmir	7.4

2.2.7 Earthquake monitoring

Under the Ministry of Earth Science (MoES), Government of India, India Meteorological Department (IMD) is the nodal agency for monitoring seismic activity in and around the country. IMD has set up a number of seismological observatories in the country for the monitoring of earthquake and first seismological observatory of the country was set up by the department at Kolkata in 1898. A real-time seismic monitoring network consisting of seventeen broad-band stations for monitoring of earthquakes is set up to estimate earthquake parameters within shortest possible time. Significant earthquakes are being auto-located and first information is sent within 15 minutes using both Indian and global seismic stations configured in the network. A 20–station VSAT-based seismic telemetry network is established for real-time monitoring and reporting of seismic activity in the northeastern region of India. The operational task of the department is to quickly estimate the earthquake source parameters immediately on occurrence of an earthquake and disseminate the information to all the user agencies including the concerned state and central Government agencies responsible for carrying out relief and rehabilitation measures.

2.2.8 Earthquake Prediction

The National Centre for Seismology (NCS) is set up to predict of earthquake by using the data sets generated by various seismic and GPS networks. The Multi-Parametric Geophysical Observatories (MPGOs) have been set up at Ghuttu, Shillong and Koyna to monitor various earthquake precursory phenomena such as seismicity patterns, crustal deformations, gravity anomalies, electrical resistivity changes, electromagnetic perturbations, water level changes, geo-hydrochemical changes, Radon and Helium anomalies and thermal anomalies, etc. Preliminary analyses of these data sets have provided useful leads on the ongoing tectonic processes in the Koyna-Warna region, which has helped in issuing short-term forecasts of earthquakes in this region. In Ghuttu, various geophysical data series are being analysed for probable identification of precursory signals.

2.3 Tsunami

Earthquakes that occur under the oceans water, caused huge waves called tsunamis. Tsunamis are long-period waves generated by disturbances, such as earthquakes, volcanic activity, and undersea landslides. The term Tsunami has been derived from a Japanese term Tsu meaning 'harbor' and nami meaning 'waves'. Tsunamis are popularly called tidal waves.

Figure 2.8 : Loss of life and property due to Tsunami
[*Source :* Times of India]

These waves, which often affect seashores, originate by rapid displacement of water from the sea either by seismic activity, landslides, volcanic eruptions or large meteoroid impacts. The crests of these waves can exceed heights of 25 meters on reaching shallow water (Figure 2.9). The unique characteristics of tsunamis (wavelengths commonly exceeding 100 km, deep-ocean velocities of up to 700 km/hour, and small crest heights in deep water) make their detection and monitoring difficult. Characteristics of coastal flooding caused by tsunamis and caused by the storm surges. The effect of a tsunami is either unnoticeable or even very destructive (Table 2.9).

Table 2.9 : History of tsunami's in India

Date	Location	Impact
1524	Near Dabhol, Maharashtra	Sufficient data not available
02 April 1762	Arakan Coast, Myanmar	Sufficient data not available
16 June 1819	Rann of Kachchh, Gujarat	Sufficient data not available
31 Oct. 1847	Great Nicobar Island	Sufficient data not available
31 Dec. 1881	Chennai	
26 Aug. 1883	Chennai	
26 June 1941	Andaman archipelago	
27 Nov. 1945	West coast of India from north of Karachi was felt at Kandla	
26 December 2004	Tamil Nadu, Kerala, Andhra Pradesh, Andaman	The East coast of India was affected.

Figure 2.9 : High waves created by Tsunami [*Source:* MoES, IMD, reports]

An earthquake-tsunami combined catastrophe is among the most widely feared natural disasters today. Any place that has experienced such a terrifying event sustains heavy damage and suffers from a massive loss of life (Figure 2.9).

2.3.1 Causes of Tsunami

The plate movements or fault movements on the seafloor and volcano caused tsunamis. During these movements, its released a huge amount of energy, having the capacity to cross the effect from oceans to lands. The second most common cause of the tsunami is a landslide either occurring underwater or originating above the sea and then plunging into the water. The largest tsunami ever produced by a landslide was in Lituya Bay, at Alaska during 1958. The massive rock slide produced a wave that reached a high watermark of 50–150 meters above the shoreline. Thus, there are three main causes of tsunami (i) Movement (ii) Landslide (iii) volcanic eruptions.

Tsunamis have occurred in all the oceans and in the Mediterranean Sea, but the great majority of them have occurred in the Pacific Ocean. Till now scientists have not developed any exact theory to predict earthquakes, therefore the prediction of exact occurrence of tsunami also not possible.

The tsunami of 2004 in India, created devastating impact on life and killed nearly 16,500 people destroyed 1.5 lakh houses and affected 18 million people.

2.3.2 Typical Adverse Effects

It is normally the flooding effect of the tsunami that causes major destruction to the human settlements, roads and infrastructure (Figure 2.10) thereby disrupting the normal functioning of the society. Withdrawal of the tsunami causes major damage. As the waves withdraw towards the ocean they sweep out the foundations of the buildings, the beaches get destroyed and the houses carried out into the sea.

Figure 2.10 : Flooding caused by the 2004 Tsunami in Tamil Nadu

2.3.3 Detection and Monitoring of Tsunamis

Deep ocean tide gauge or buoys are used to confirm the occurrence of tsunami waves in India. These buoys observe and record changes in sea level in deep ocean. Six deep-ocean assessment and reporting of tsunami DART buoys and 25 seismographic stations relaying information to 26 national tsunami information centres. In addition to buoys, the Survey of India maintains a tide gauge network along the coast of India. The gauges are located in major ports. Apart from the tide gauge, a tsunami can be detected with the help of radars.

Tsunami warning system (TWS)

A network of sensors is established to detect tsunami, communication and to issue timely alarms to permit evacuation at the coastal area. There are two distinct types of tsunami warning:

a) International tsunami warning systems and

b) Regional warning systems.

Ministry of Earth Sciences has set up an Indian Tsunami Early Warning Centre at the Indian National Centre for Ocean Information Services (INCOIS), Hyderabad on 26th December, 2004. INCOIS has a data warehouse of ocean related information gathered from various institutions and sources in India. The Centre is mandated to provide advance warnings on Tsunamis likely to affect the coastal areas of the country. India Meteorological Department has been set up a 17–station Real-Time Seismic Monitoring Network (RTSMN). The network is capable of monitoring and reporting, in least possible time, and to record the occurrence of earthquakes capable of generating Tsunamis likely to affect the Indian coastal regions and issue advisory. The data from the 17 Broadband seismic field stations is transmitted, simultaneously in real-time through V-SAT communication facilities to the Central Receiving Stations (CRSs) located at IMD, New Delhi and INCOIS, Hyderabad for processing and interpretation.

2.4 Volcanoes

Volcanoes are a natural hazard and create hot gases and ash, which are falling on the earth surface. There are cracks/holes in the earth's crust through which molten rock and gases come out on the earth surface and this is known as volcanoes. Volcanic eruptions include lava flows, falling ash, mudflows, and toxic gases. Volcanic activity may also trigger other natural hazardous events including local tsunamis, landslide and floods.

There are two classes of eruption during volcanic activity:

Explosive eruptions: In this eruption rapid dissolution and expansion of gas from the molten rock comes out beneath the earth's surface(Figure 2.11).

Effusive eruptions: In this eruption materials flow rather than explosions on the earth surface from multiple points. Flows are governed by gravity, surrounding topography, and material viscosity (Figure 2.12).

Figure 2.11 : Barren Island of India eruption during 2017 and release of gas [*Source:* Online search free download image]

Figure 2.12 : Release of molten magma and materials due to volcanic eruption [*Source:* Online search free download image]

2.4.1 Causes of Volcano

Earth's upper mantle or lower crust melts and discharged through an opening or vent due to pressure difference. Volcanic activity frequently occurs at the boundaries of earth's tectonic plates. Earthquake activity around a volcano can provide valuable information. Earthquakes also can indicate that the magma is moving beneath a volcano. However, eruptions can occur with no perceivable change in seismic activity.

2.4.2 Volcano Prediction

It is difficult to predict volcanic eruptions. Seismographs/seismometers EDM levelling, GPS is used to pinpoint the earthquakes, which tract the rise of magma and its movement along fissures. Measurements of the tilt by tiltmeter of entire mountain provide additional information about the breathing of the volcano. Instruments that sniff SO_2, CO_2 and other gases also can pass signal changes in volcano. Infrared camera, satellite images, spectrometer, satellite interferometer are also used for volcano monitoring and prediction.

2.4.3 Major Volcano Events in India

A conventional way of classifying magmatic volcanoes is by their frequency of eruption. They are called active if they erupt regularly. Dormant or inactive volcanoes are those that have erupted in the past times, but are now quiet, while the volcanoes that have not erupted since ages are termed extinct. India has a list of six/seven volcanoes (Table 2.10) including only the live one, Barren Islands volcano (situated in Indian Union Territory of Andaman and Nicobar Islands and lies about 138 km northeast of the Port Blair) which also happens to be the only confirmed active volcano in South Asia. Barren Island, one of the most easterly of the Andaman Islands, is the only confirmed active volcano in India.

The first recorded volcanic eruption in India dates back to 1787. The Barren Island volcano, in the Andaman and Nicobar Islands is erupting once again. After remaining dormant for 150 years, it first erupted in 1991 and now showed signs of powerful volcanic eruptions. The first eruption of the volcano was recorded in the year 1787, since then it erupted 11 times with the most recent one being on January 23, 2017.

Table 2.10: Active and extinct volcanoes of India

S.No.	Name	State	Type/Status	Last eruption
1	Barren Island	Andaman Islands	Strato volcano/Active	Active 2017
2	Narcondam Island	Andaman Islands	Strato volcano/dormant	
3	Baratang	Andaman Islands	Mud volcano/Dormant	Active since 2003
4	Deccan Traps	Maharashtra		
5	Dhinodhar Hills	Gujarat	Extinct	Inactive volcano
6	Dhosi Hill	Haryana	Extinct	
7	Tosham Hill	Haryana	Extinct	

2.5 Cyclone

The cyclone frequently affects the Indian subcontinent and the parts of the world. India's long coastline of 7,516 kilometer and about 8 per cent of the land is vulnerable to cyclones. On an average, five to six tropical cyclones every year are observed over the Indian subcontinent. Cyclone occurs frequently on both the coasts and is exposed to nearly 10 per cent of the world's tropical cyclones. More cyclones occur in the Bay of Bengal than in the Arabian Sea. In India, Tropical cyclones occur in the months of May–June (pre monsoon) and October–November (post monsoon).

Cyclone is a type of storm and is a region of intense low atmospheric pressure (Figure 2.13) surrounded by high atmospheric pressure accompanied by very high winds blowing in anticlockwise direction in the northern hemisphere and in the clockwise direction in the southern hemisphere. It is accompanied by very high wind speed (strong gales) and lashing rain and creates tidal waves that cause floods in coastal areas. A cyclone can carry the power to destroy buildings, blow away a roof, uproot trees, crops and wash away homes. Storm surges are the greatest killers generated by a cyclone and seawater inundates low lying areas of coastal regions, which causes heavy floods, erodes beaches and embankments, destroys vegetation and reduces soil fertility. They occur mainly in the tropical and temperate regions of the world. Cyclones in India are generally moderate in nature.

Figure 2.13 : Cyclone hitting Tamil Nadu and Andhra Pradesh states of India [*Source:* MoES, IMD, Webpage]

Cyclones are known by different names in different parts of the world and also regional names are given below:

- Typhoons in the Northwest Pacific Ocean, Sea off China, Philippines, Japan, Islands of Western pacific.
- Hurricanes in the North Atlantic Ocean, the Northeast Pacific Ocean, or the South Pacific Ocean and pacific coast of Maxico.
- Tropical cyclones – the South-west Pacific Ocean and Southeast Indian Ocean.
- Severe cyclonic storm" (the North Indian Ocean)
- Tropical cyclone (the Southwest Indian Ocean)
- Willie-Willie in Australia, East Coast of Africa and North-west and North-east coast of Australia.
- Tornado in South America.

2.5.1 General Characteristics

Some of the general characteristics of a cyclone are:

1. Strong winds
2. Exceptional rain
3. Storm surge

2.5.2 Factor Responsible for Cyclone

Tropical cyclones are originate over the warm tropical oceans, where the surface temperature is about 27°C or above and require a pre-existing mild tropical disturbances, which under favourable condition intensify and develop into a cyclone. There are certain basic requirements described below are to be fulfilled for the development of cyclone:

- Large and continuous supply of warm and moist air
- Large values of Coriolis force
- Existence of weak tropical disturbances
- Upper air outflow
- Weak vertical wind shear in the basic current
- Small atmospheric vortices in intertropical convergence zone.

2.5.3 Distributional Pattern of Indian Cyclones

About two-thirds of the cyclones that occur in the Indian coastline occur in the Bay of Bengal (Figure 2.14). The states, which are generally affected in the east coast are West-Bengal, Odisha, Andhra Pradesh, Tamil Nadu and on the west coast Gujarat, Maharashtra, Goa, Karnataka and Kerala. Cyclones typically strike the East Coast of India, along the Bay of Bengal, *i.e.,* the states of West Bengal, Odisha, Andhra Pradesh and Tamil Nadu, but also parts of Maharashtra and Gujarat at the Arabian Sea West Coast.

The areas generally affected by cyclones in India are shown in Figure 2.15. Some of the major cyclones that have affected the country in the past are as mentioned in Table 2.11.

Table 2.11 : Death associate with noteworthy Tropical Cyclones (1970 – 2005)

S.No.	Year	Area	Death toll
1	1971	Eastern Coast	9658
2	1972	Andhra Pradesh and Odisha	100
3	1977	Chennai, Kerala and Andhra Pradesh	14,204
4	1979	Andhra Pradesh	594
5	1981	Gujarat	470
6	1982	Gujarat and Maharashtra	500
7	1984	Tamil Nadu and Andhra Pradesh	512
8	1985	Andhra Pradesh	5000
9	1990	Andhra Pradesh	957

S.No.	Year	Area	Death toll
10	1990	Odisha	250
11	1999	Odisha	8913

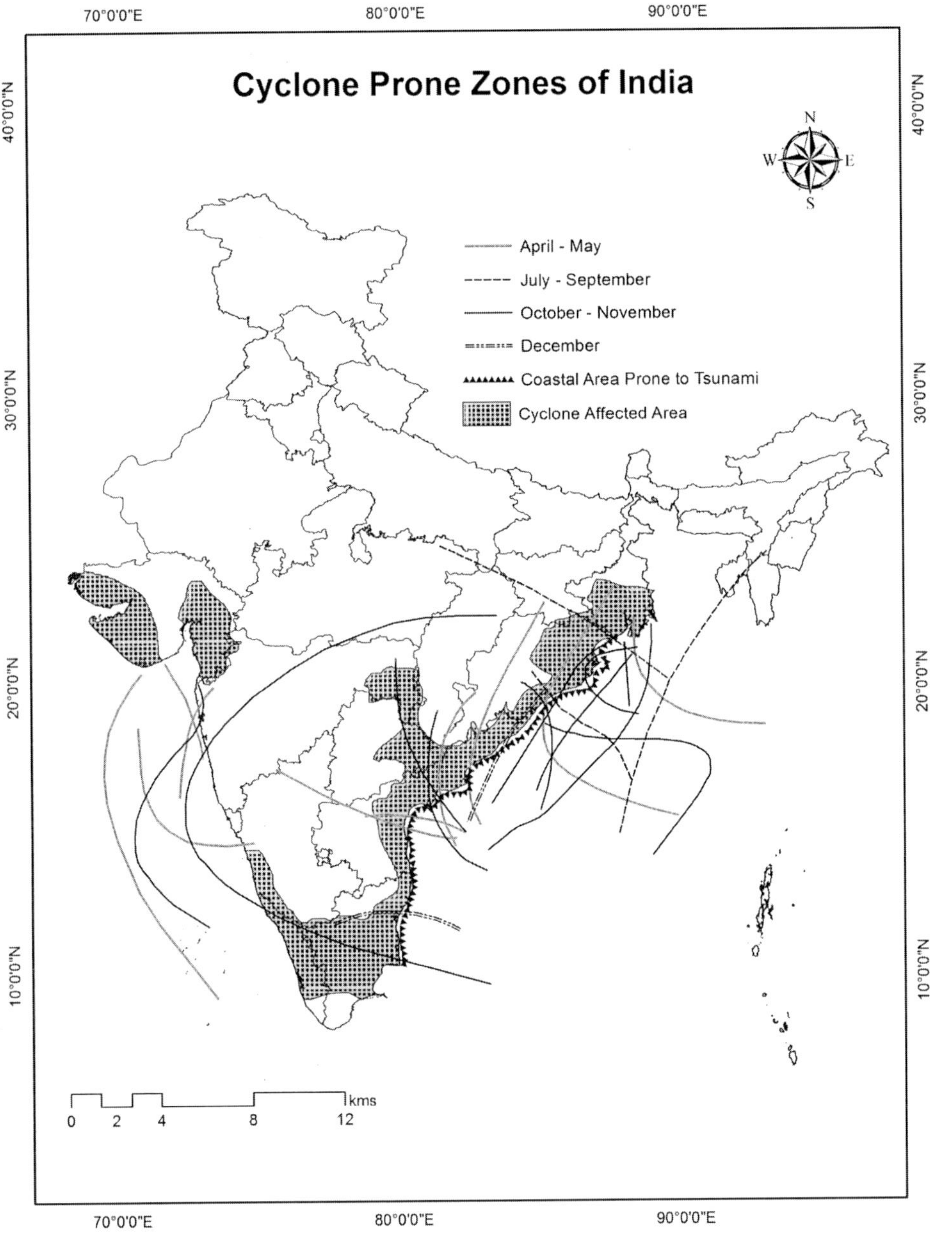

Figure 2.14 : Indian coastal area affected by cyclone (*Source:* MoES, IMD)

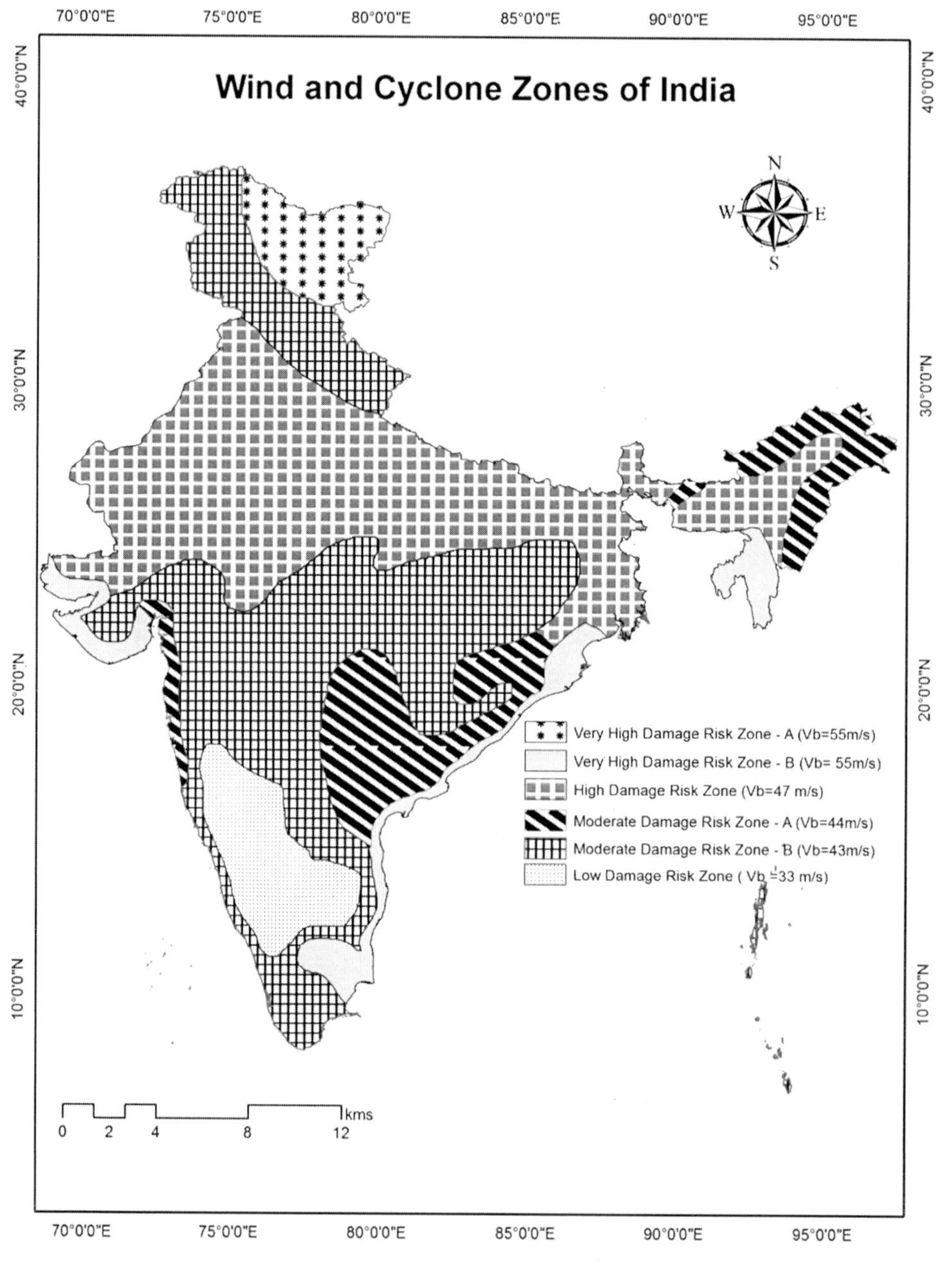

Figure 2.15 : Wind and Cyclone map of India (*Source:* MoES, IMD)

2.5.4 Odisha Super Cyclone

On 29th October 1999, a super-cyclone with wind speed of 260–300 km/hour hit the 140 kilometer coast of Odisha and created a storm surge in the Bay-of-Bengal with water level 9 meters higher than normal. The superstorm travelled more than 250 km inland and killed thousands and devastated millions. It had caused 10,000 death and destroyed 2.75 lakh houses. The another major cyclone *Aila*, which occurred in the year 2009 caused 120 deaths and affected 5.4 million people.

2.5.5 Cyclonic Storm Nilam

A tropical storm on 1st November, 2012 hit southern Indian coast and due to heavy rainfall and storm surge flooding in low-lying area was reported (Figure 2.16). This storm known as *Nilam* and was originated from an area of low pressure over the Bay of Bengal on October 28, and within two days, this depression gradually intensified into a deep depression, and subsequently a Cyclonic Storm by October 30. This storm had displaced more than 1 lakh people and killed 6 person.

Figure 2.16 : Cyclone 'Nilam' on November 1, 2012 [*Source:* MoES, IMD]

2.5.6 The Naming of Tropical Cyclones

After 1978, World Meteorological Organization (WMO) has suggested to name the storms and started giving an alphabetical list of names of storms in its annual meetings. These lists are rotated every six years. If a named cyclone has occurred than its name may be deleted from the listed names of the cyclone.

After WMO initiatives, in India, Indian Meteorological Department (IMD) is naming the storms (tropical cyclones) to identify these storms, so that people could be informed about their arrival quickly. This naming of cyclones that form over the North Indian Ocean region, was taken by unanimously decision taken in the year 2000 by eight countries, *i.e.,* Bangladesh, India, Maldives, Myanmar, Oman, Pakistan, Sri Lanka and Thailand.

Each country gave eight names for the cyclones. Thus a list of cyclone names was prepared. It was also decided that the 8 countries will take turns to name of the cyclones. IMD has a list of 64 names contributed by eight countries of the Indian sub-continent— Bangladesh, India, Maldives, Myanmar, Oman, Pakistan, Thailand and Sri Lanka.

Tropical cyclones in the Bay of Bengal and the Arabian Sea are being named since 2004 (Table 2.12). The cyclonic storm Nilam that has hit the eastern coast of India was the name given as 'Nilam' by Pakistan as per the cyclone naming procedure. Some of the names given by different countries are shown in table 2.13. A list of some of the deadly hit cyclones of the world is shown in table 2.14.

Table 2.12 : Name of Cyclones and affected states of India after 2004

Affected state	Lowest pressure of cyclone (mb)	Name of the cyclone	Year of occurrence
Gujrat	990	Onil	2004
	986	Yemvin	2007
Kerla	999	Fanoos	2005
Odisha	986	Laila	2010
Tamilnadu	998	Fanoos	2005
	996	Nisha	2008
	988	Jal	2010
	972	Thane	2011
	992	Nilam	2012
Andhra Pradesh	986	Yemvin	2007
	996	Khai-Muk	2008
	986	Laila	2010
	992	Nilam	2012
Maharashtra	988	Phyan	2009
	988	Jal	2010

Table 2.13 : Names of Cyclones given by different country

S.No.	Country	Name given to cyclone	Total number of names
1	India	*Agni, Akash, Bijli, Jal, Lehar, Megh, Sagar, Vayu*	08
2	Pakistan	*Fanoos, Nargis, Laila, Nilam, Nilofar*	05
3	Sri Lanka	*Mala, Rashmi, Bandu, Priya*	04
4	Bangladesh	*Onil, Ogni, Nisha, Giri*	04
5	Oman	*Baaz, Sidr, Murjan*	03
6	Myanmar	*Pyarr, Phyan, Thane*	03
7	Thailand	*Mukda, Phet*	02
8	Maldives	*Goni*	01
		Total	30

Table 2.14 : The deadly Storms that hit world wide and number of death toll

Affected Area	Number of death	Name of the cyclone	Year of occurrence
Caribbea	22,000	Great Hurricane	1780
Texas, southern US	8,000	Galveston Hurricane	1900
Bangladesh	nearly Half a million	Cyclone Bhola	1970
Indian coast	15,000	Andhra Pradesh Cyclone	1977
Nicaragua,	10,000	Hurricane Mitch	1998
Odisha, India	10,000	Odisha super cyclone	1999
US.	1,800	Hurricane Katrina	2005
Myanmar	150,000	Cyclone Nargis	2008
Cuba, Haiti, Bahamas and US	200	Hurricane Ike	2008

2.5.7 Cyclone Monitoring and Forecasting

Indian Meteorological Department has setup the network of Radar over the east and west coast of the country to monitor the cyclone. Through these radars, a constant watch is kept for the likely genesis of a tropical cyclone with the satellite imagery, data from ships and ocean buoys are also very valuable. When the system comes nearer to the Indian coastline, subsequent development and movement are monitored by chain of cyclone detection radar. The likely movement of storm is predicted with the help of tract prediction models along with numerical weather models and by reference of the past climatology of cyclone data.

2.6 Storm Surge

A storm surge is a situation of the abnormal sea level rise caused by strong atmospheric disturbance (such as strong winds and low atmospheric pressure) near the centre of cyclone or hurricanes. The sudden and sharp rise of coastal sea water caused by their combination often results in severe surge disaster. Sometime, it is even more severe than the earthquake with regard to death and destruction.

2.7 Cloud Burst

A cloud burst represents highly concentrated rainfall over a small area lasting for a few hours. Any sudden and heavy precipitation that takes place due to convective activity and the occurrence of thunderstorms may be classified as cloudburst. A cloudburst is an extreme form of rainfall, sometimes mixed with hail and thunder, is capable of creating flash flood conditions. The occurrence of cloudburst restricted during the monsoon period and occurs frequently during July–October in the Himalayan region of northern India. It occurs, when a pregnant monsoon cloud drifts northwards, from the Bay of Bengal across the Ganges plains, then onto the Himalayas and bursts, bringing rainfall as high as 75 millimeter/hour; such situations are created by westward passage of monsoon depression and mid-troposphere low-pressure system. They lead to flash floods (a flood, which occurs suddenly).

2.7.1 Cloud Burst Events in India

Cloudbursts lead to a distressing situation. A cloud burst over New Delhi, between 20th and 21st July, 1958 provided 266 mm of rainfall in a 24 hours. About 188mm (68 percent) of rainfall in 3 hours was towards the early parts of the morning of July 21. The rate of rainfall was over 75mm/hour.

A record rainfall of 294 mm was recorded at Ajmer on July, 16 in 1979. Apart from extensive inundation, a railway bridge over the Luni river was washed away. Some important events of cloudbursts are given in Table 2.15:

Table 2.15 : Some events of cloud burst in India

S.No.	Place	Date	Impact
1	Kemra, Kothiara, Tehri, Uttarakhand	28 May, 2016	120 houses, 100 animals
2	Tehri district, Uttarakhand	May 28, 2015	At least 100 animals were buried under the debris although there were no human casualties
3	Sonamarg, Srinagar, Jammu and Kashmir	16 Jul 2015	At least 4 humans died.
4	Shimla, Himachal Pradesh	14 May, 2015	Four youths went missing after cloud burst triggered a series of flash floods

S.No.	Place	Date	Impact
5	Kashmir Valley	6 Sep., 2014	200 people died
6	Kedarnath, Uttarakhand	15 Jun., 2013	1000 people died, 400 insured and 365 houses damaged
7	Pandrasu ridge, Garhwal, Himalaya Uttarakhand	3 Aug., 2012	35 people died and 436 livestock lost
8	Leh, Jammu and Kasmir	6 Aug., 2010	255 people died and 1749 houses damaged
9	Pithoragarh, Uttarakhand	7 Aug., 2009	38 people died
10	Bhavi, Himachal Pradesh	14 Aug., 2007	52 people died
11	Mumbai	26 July, 2005	1000 people died
12	Shillagarh, Kullu, Himachal Pradesh	16 July, 2003	40 humans died
13	Gona, Tehri, Uttarrakhand	31 Aug., 2001	22 7 humans, 7 animals, 28 houses
14	Malpa- Pithoragarh, Uttarakhand	17 Aug., 1998	250 humans, 60 animals, 40 houses
15	Okhimath-Rudraprayag, Uttarakhand	11 to 19 Aug., 1998	103 humans, 422 animals, 820 houses
16	Chirgaon, Shimla	15 Aug., 1997	1500 human died
17	Nilkanth-Puri Garhwal, Uttarakhand,	09 July, 1990	100 humans, 10 houses
18	Karmi-Kapkote, Bageshwar, Uttarakhand	22 July, 1983	150 humans, 20 animals, 6 houses
19	Kuntha-Rudraprayag, Uttarakhand	17 Aug., 1979	39 humans, 39 animals
20	Ganai-Almora, Uttarakhand,	20 July, 1970	200 humans, 33 houses
21	Delhi	21 July, 1958	Many parts of the city inundated

Cloudburst at Ghansal Tehsil, Uttaranchal

Heavy casualties resulting in the death of 28 persons and many lost of cattle were reported in the village Marwari, Medha, Agunda, Kol and Manijara villages on 10th Aug., 2002 at Tehsil Ghanasli district Tehri Garwal due to cloudburst and landslides.

The houses and cultivated fields just nearby this channel marwari nala (small drainage channel) was damaged and washed out. The obstruction of natural flow led to the collapse of said house and cultivated fields in Medha villages also. The Kol village is situated on an old landslide zone (residual deposits). Three persons and many cattle were reported to have been killed in a house collapse apart from damage to the cultivated fields and other houses.

During this event, huge landslides, bring down boulders and uproots trees. There is another type of flood, the flash flood can occur in smaller rivers and streams.

Types of Damage and Loss

Damage to agricultural land	-	407 hectares.
Number of village affected	-	17
Number of causalities	-	28
Number of affected families	-	228
Loss of livestock	-	Dead 61, Missing -38
Damage to houses	-	38 Fully, 151 Partially

Cloudbursts descend from very high clouds, sometimes with tops above 15 kilometers and rain from a cloudburst is usually of the shower type with a fall rate equal to or greater than 100 mm per hour (Figure 2.17).

Figure 2.17 : Cloud burst image

On 5th August, 2010, sudden overnight rains due to cloud burst caused flash floods in the town of Leh, in northern Ladakh region that borders China and Pakistan, killing more than 100 people and leaving hundreds injured (Figure 2.18).

Figure 2.18 : Cloud burst in Ladakh [*Source:* Online search free download image]

In 2010 Cloudburst led flash mudslides and flash floods killed 196 people, including six foreigners and injured more than 400 and swept away the number of houses, sweeping away buildings, bus stand and military installations in trans Himalaya Leh town of Jammu and Kashmir. In June 2013, a multi-day cloudburst centred on the North Indian state of Uttarakhand caused devastating floods and landslides in the country's worst natural disaster. More than 5,700 people were dead, destruction of bridges and roads left about 100,000 people alone.

2.7.2 Cloudburst Prediction

At present, it is difficult to predict cloudburst with a more lead time. The cloudbursts are predicted two to three hours in advance in terms of heavy rainfall using Doppler Weather Radar (DWR) and Satellite Data. Its formation and development is very fast and happens in very small locality and hence it is difficult to forecast. As per Ministry of Earth Science (MoES) statement; India will be able to forecast cloud bursts at least two days in advance from 2022 onward.

Cloud bursts are extreme weather events in which an area registers more than 100 mm rain in just one hour. At present India's weather forecasting system depends on high-performance computing (HPC) system with a capacity of 10 PetaFlops with a resolution to 12 km. With advanced computing system MoES will be able to give forecast up to 5 km, or maybe even up to 3 km, and able to forecast cloud bursts at least two days in advance by 2022.

2.8 Landslide

A landslide is a geological phenomenon, which includes a wide range of rock/land movement, such as rockfalls, deep failure of slopes, and shallow debris flow. It includes all varieties of mass movements of hill slopes and can be defined as the downward and outward movement of slope forming materials composed of rocks, soils, artificial fills or includes slides, falls and flows of unconsolidated materials.

2.8.1 Causes of Landslide

Landslides can be triggered by earthquakes, volcanic eruptions, soil saturated by heavy rains or groundwater rise, and river undercutting. Earthquake shaking of saturated soils creates particularly dangerous conditions. Although landslides are highly localized, they can be particularly hazardous due to their frequency of occurrence.

There are several causes of the landslide. Some of the major causes are as follows:

- Weakness in the composition and structure of rock or soil
- Erosion
- Intense or prolonged heavy rainfall
- Human Excavation
- Vibration caused by Earthquake
- Volcanic eruption
- Cloud burst
- Undercutting of cliffs and banks by waves or river erosion
- Removal of vegetation and toe erosion.

2.8.2 Distributional Pattern

Landslides constitute a major natural hazard in our country, which accounts for considerable loss of life and damage to communication routes, human settlements, agricultural fields and forest lands. The Indian subcontinent, with diverse physiographic, seismic, tectonic and climatological conditions, is subjected to a varying degree of landslide hazards; the Himalayas including North-eastern mountains ranges being the worst affected, followed by a section of Western Ghats and the Vindhyas.

Torrential rainfall on the deforested slopes is the main factor in the Peninsular India.

2.8.3 Landslide Affected Region of India

In the hilly terrain of India including the Himalayas, landslides have been a major and widely spread natural disasters that often strike life and property. Landslides occur in the hilly regions such as the Himalayas, North-East India, the Nilgiris, and Eastern and Western Ghats). Landslide-prone areas largely correspond to earthquake-prone areas, *i.e.,* North-West and North-East, where the incidence of landslides is the highest (Figure 2.19).

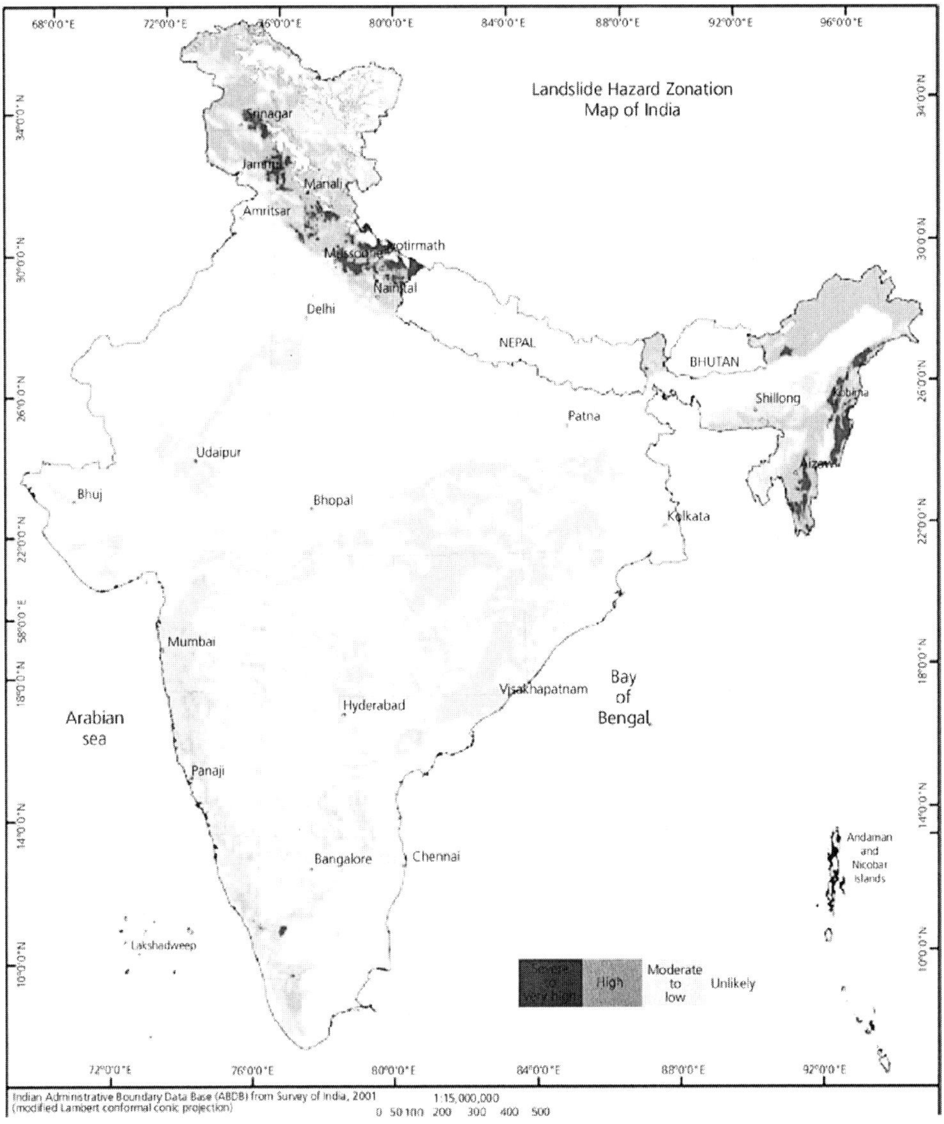

Figure 2.19 : Landslide hazard zone map of India (Dark areas more susceptible than light yellow areas). (*Source :* NDMA, Report)

2.8.4 Landslides and Loss of Life

Landslides of large dimension are still active and found in the mountains regions. They are most often found in valleys and valley basins, where decomposed tuffs and other unconsolidated sediments are abundant near the surface. This is even more typical for landslides in hill regions and along bluffs. The reactivation of landslides may be related to the irregular precipitation distribution under continental climate, to years of wet weather and even more characteristically to heavy rainfall in extreme weather. Particularly for shallow landslides, exceptionally high precipitation and the reactivation of landslides are strongly correlated. One of the worst tragedies took place at Malpa (Uttarkhand) on 11th and 17th August, 1998 when nearly 380 people were killed and massive landslides washed away the entire village. A list of major landslide events in India is given below Table 2.16.

Table 2.16 : List of Major landslides events in India

Place	Date	Impact
Guwahati, Assam	18 September, 1948	500 people died
Darjeeling	04 October, 1968	More than 1000 people died
Malpa, U.K.	11–17 August, 1998	More than 380 people died
Mumbai	12 July, 2000	78 people killed
Kerala	09 November, 2001	40 people died
Kedarnath, U.K.	15–16 June, 2013	5700 people died
Malin, Maharashtra	30 July, 2014	136 people died

2.8.5 Adverse Effects

The most common elements at risk are the settlements built on the steep slopes, built at the toe and those built at the mouth of the streams emerging from the mountain valley. All those buildings constructed without appropriate foundation with weak construction is venerable to landslide (Figure 2.20).

Figure 2.20 : Landslide in hilly terrain of India
[*Source:* Online search free download image]

2.8.6 Forecasting of Landslide

Early warning system based on rainfall threshold data and landslide records could be used for forecasting of landslide in India. National centre for medium range weather forecasting is also developing a complex high-resolution numerical weather model to enhance forecasting of a landslide in vulnerable area.

2.9 Avalanches

Avalanches are one of the biggest hazards in the hilly area and mountainous regions for both life and property. The forces generated by avalanches can damage or destroy most man-made structures, crops, animals and human life. An avalanche is caused, when a build-up of snow is released down a slope, and is one of the major dangers faced in the mountains in winter. An avalanche is a type of gravity current (Figure 2.21).

Avalanches are river like the speedy flow of snow or ice descending from the mountain tops. On global scale, the countries like India, Nepal and Bhutan has one of the highest incidences of avalanche and landslides. In Himalayas, avalanches are common in Drass, Pir Panijat, Lahaul-Spiti and Badrinath areas.

Figure 2.21 : Avalanche in Ladhakh India (*Source:* Online news Mumbai)

2.9.1 Area Prone to Avalanches

The topography of India has related to avalanches activity and a topographic map of India is shown in Figure 2.22. Avalanches are common in the Himalayan region. The high Himalayan regions above 3500 meter elevation are more prone to avalanches. The other factors affecting the avalanche events are given below:

- Very frequent on slopes of 30–45°.
- Convex slopes more prone to this disaster.
- North facing slope have avalanches in winter and south facing slopes during spring.
- Slopes covered with grass more prone to this hazard.
- Avalanches: They are the rapid downslope movement of snow, ice, and associated debris such as rocks and earth material.

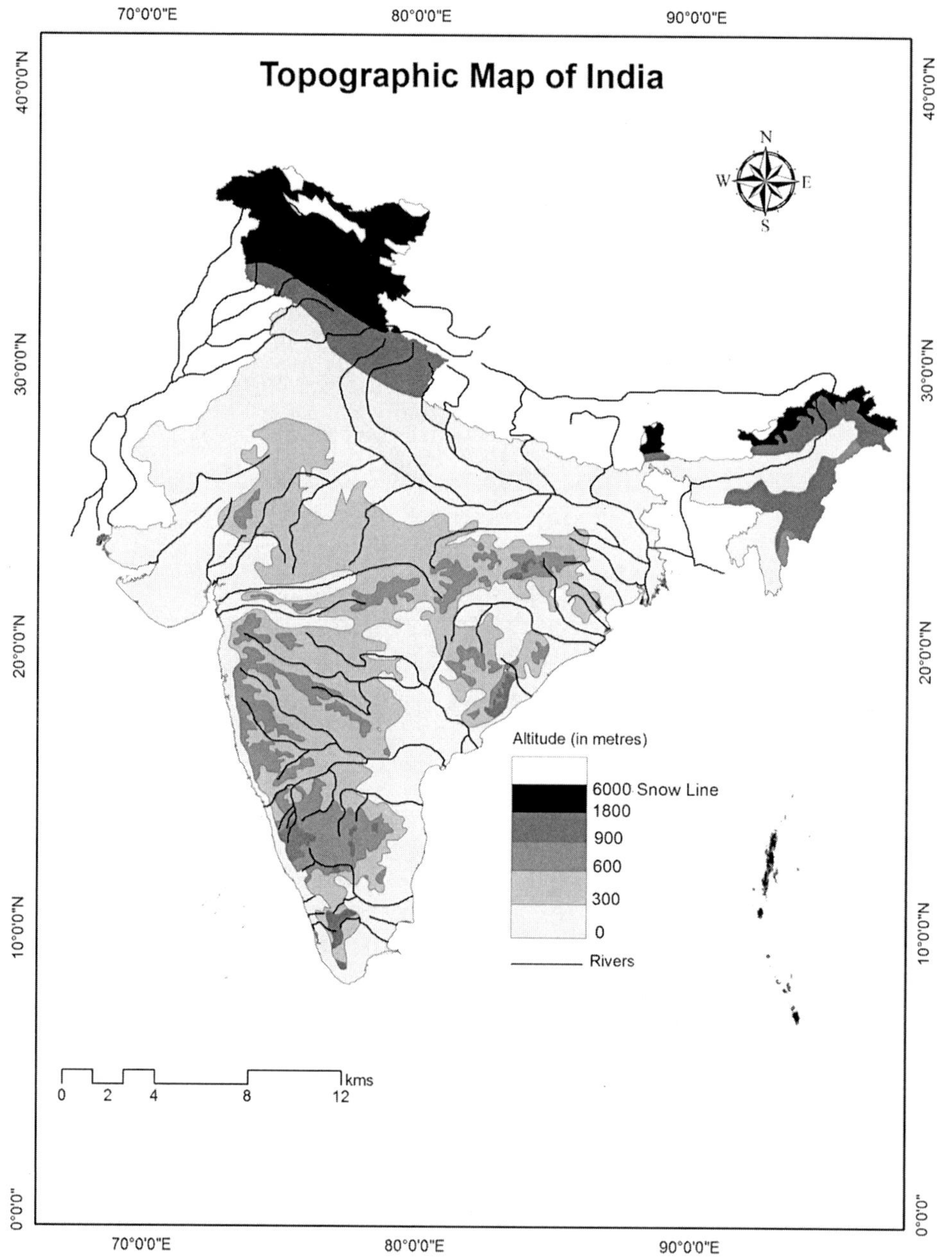

Figure 2.22 : Map showing topographical variation in India

There are three types of avalanches:

i) Point release avalanches

It occurs, when the weight of the snowpack exceeds the shear strength within it and is very common on steeper terrain. Generally, it forms a teardrop appearance.

ii) Slab avalanches

It occurs, when a large slab of snow breaks and starts moving downward at high speed. It is very fatal and accounts for around 90 per cent of avalanche related losses.

iii) Slush avalanches

It occurs, when the saturated (by water) snowpack starts and spread out from a point. The avalanches moving downwards along the slope, accumulate snow and grow in size. When the snow mixed with the air it will formed a powdery cloud.

2.9.2 Avalanche Events in India

Some of the prominent avalanche events occurred in India are given below.

- An avalanche has occurred on 18 February, 2007, in Sikkim and three persons died, many missed.
- An avalanche has occurred on 25 September 2007 in Sonmarg area of Srinagar and one person died and damaged other property.

The avalanches are mostly confined from Jammu-Kashmir, Uttarakhand, to Himachal Pradesh in high Himalayan region. Some of the avalanches events are shown in Table 2.17.

Table 2.17 : List of some avalanche events in India

S.No.	Place	Date	Impact
1	Kashmir	13 January, 2020	At least six Indian soldiers and six civilians were killed in five separate avalanches in the Himalayan region of Kashmir.
2	Kashmir	18 Nov., 2019	The event has hit one of its posts on a Himalayan glacier in disputed Kashmir region, killing 4 soldiers and 2 civilian porters.
3	Himachal Pradesh	23 and 20 Feb., 2019	Five Army jawans, who were trapped in an avalanche that struck near Shipki La on Sino-India border in Himachal Pradesh. Six jawans of the Army's 7 J & K Rifles — four from Himachal Pradesh, one each from Uttarakhand and Jammu and Kashmir were buried under the avalanche that occurred near the Shipki La border outpost in Himachal Pradesh.

S.No.	Place	Date	Impact
4	Jammu and Kashmir	25 January, 2017	The avalanche happened at height of more than 5,485 meters in the northern part of the Siachen Glacier disputed Kashmir region, killing four soldiers and their two civilian porters.
5	Jammu and Kashmir	February 3, 2016	The avalanche struck the Indian military base located in the northern Siachen Glacier, at a height of 19,600 feet and had trapped 10 soldiers under deep snow.
8	Jammu and Kashmir	April 25, 2015	At least 18 people die and dozens were injured in an avalanche triggered by a magnitude-7.8 earthquake in central Nepal. It is the deadliest avalanche ever recorded on Mount Everest.
9	Jammu and Kashmir	April 18, 2014	At least 16 Sherpa guides died in an avalanche. They were fixing ropes for climbers at an elevation of 19,000 feet when the avalanche hit.
10	Gayari Sector avalanche	2012	Around 129 soldiers and 11 civilians were dead.
11	Himachal Pradesh	March, 2011	Near Pindri Nala 2 labourers died.
12	Himachal Pradesh	March, 1991	Tinku avalanche occurs every year 4–5 times from Jan to March. Road was blocked for 40 days in 1991
13	Himachal Pradesh	March, 1979	About 237 people killed. Communication disrupted
14	Lahaul Valley avalanche	1979	About 200 people killed. Communication disrupted
15	Lahaul and Spiti	March, 1978	About 30 people killed, road and property damaged.
16	Lahaul and Spiti	Jan., 1975	Earthquake shocks triggered the avalanche of great dimensions damaging road network.

2.9.3 Forecasting of Avalanche

India has developed an avalanche-forecasting technique using near real-time remote-sensing data and field-based observation from several observatories and automatic weather stations. It uses the MM5 weather-forecasting model, which provides information on the volume of snow, altitude, and catchment areas, based on which the forecast is made. Doppler Weather Radars with advanced polarimetric capability is deployed in the Himalayan region to monitor and detection of clear air turbulence and tracking of cloud burst, hailstorm, landslides and other severe weather events.

The Hidden Markov Model (HMV) is used for avalanche forecasting in Jammu and Kashmir and using mesoscale weather forecasting model (MM5) has been used for its forecasting in Indian Himalaya.

2.10 Flood

A flood is a temporary inundation of dry land with water. It is caused by overflowing of rivers, heavy precipitation, cloud burst, storm surge, tsunami, excess runoff etc. (Figure 2.23). Flood is a state of high water level along a river channel or on the coast that leads to inundation of land, which is not usually submerged.

India rainfall does not occur throughout the year, 75 per cent of the annual rainfall is received during monsoon (June- September). Heavy rainfall or prolonged downpour often causes floods in the low-lying coastal areas and over-flows lakes and rivers resulting in floods. Floods have been regular features of some parts of India causing huge economic loss as well as a loss of life. The area vulnerable to floods is 40 million hectares and the average area affected by floods annually is about 8 million hectares. About 30 million people are affected by flood every year. Floods in the Indo-Gangetic-Brahmaputra plains are an annual feature. On average, a few hundred lives are lost, millions are rendered homeless and several hectares of crops are damaged every year. Floods are a perennial phenomenon in at least 5 states – Assam, Bihar, Odissa, Uttar Pradesh and West Bengal. On account of climate change, floods have also occurred in recent years in areas that are normally not flood prone. The drought prone parts of Rajasthan experienced floods during 2006.

Types of Floods

The common types of flood are of three types

A) Fluvial or river flood
B) Pluvial flood
C) Costal or surge flood

A) **Fluvial or river flood :** It is a flood that occurs ,when the water level in a river, lake or stream rises and overflows onto surrounding banks, shores and neighboring land due to excessive rainfall or snowmelt.

B) Pluvial flood occurs due to extreme rainfall events

There are two types of pluvial flood

1. Surface flood
2. Flash flood

Surface flood: It occurs when the urban drainage system is overwhelmed nearby structures. It is a slow or gradual type of flood than the flash flood, which may provide time to people to save themselves.

Flash flood: It is an intense, high velocity torrent of water triggered by torrential rain falling within a short period of time from nearby elevated terrain. It is mainly caused by cloud burst in hilly areas. It is very destructive type of flood.

C) Costal or surge flood: It occurs in area that lie on the coast of a sea, ocean or other large body of open water. It is a result of extreme tidal conditions caused by severe weather.

2.10.1 Flooding : Flooding is of two types:

a. Coastal flooding

Coastal flooding is caused due to storm surges, excessive run-off from the upper watershed. Flooding of deltas and low-lying coastal areas is caused by tidal waves or frequent channel shifts.

b. River flooding

This type of flooding is caused due to excessive run-off brought by heavy rains, Land-borne floods occur when the capacity of stream channels to conduct water is exceeded and water overflows banks.

Figure 2.23 : Flooding in the Indo Gangetic plain zone

Causes

There are several causes of floods and some of them listed below.

- Heavy rainfall
- Passage of a low-pressure area/Cyclone
- Heavy siltation of the river bed
- Blockage in the drains leads to flooding of the area.
- Cloud burst and landslides blocks the flow of the stream.

2.10.2 Adverse Effects

The most important consequence of floods is the loss of life and property. Structures like houses, bridges, roads etc. get damaged and there is a huge loss to life and livestock. Floods are generally affects drinking water facilities, contamination of water, outbreak of epidemics, diarrhoea, viral infection, malaria and many other infectious diseases.

Flooding also affects a large area of agricultural land and crop which are getting inundated and creates a huge crop loss. This results in shortage of food, and animal fodder. Floods may also affect the soil fertility as well as its characteristics. The land may be rendered infertile due to erosion of top layer or may turn saline if seawater floods the area.

2.10.3 Flood Warning

Flood forecasting is done in India by central groundwater board. IMD is warning about formation of low-pressure area its movement and associated amount of precipitation and their life span.

2.10.4 Floods Prone Area in India

Floods occur in almost all the river basins of the country. The flood-prone region of India shows pictorially the areas liable to floods and shown in figure 2.24. Around 12 per cent (40 million hectares) of land in India is prone to floods.

Most of the flood affected areas lie in the Ganga basin, Brahmaputra basin, the North-western river basin (comprising Jhelum, Chenab, Ravi, Sutlej, Beas and the Ghagra), Peninsular river basin (Tapti, Narmada, Mahanadi, Baitarani, Godavari, Krishna, Pennar and the Kaveri) and the coastal regions of Andhra Pradesh, Tamil Nadu, Odisha and Kerela. Assam, Uttar Pradesh, Bihar and Odisha are some of the states, which come under highly flood-prone zone.

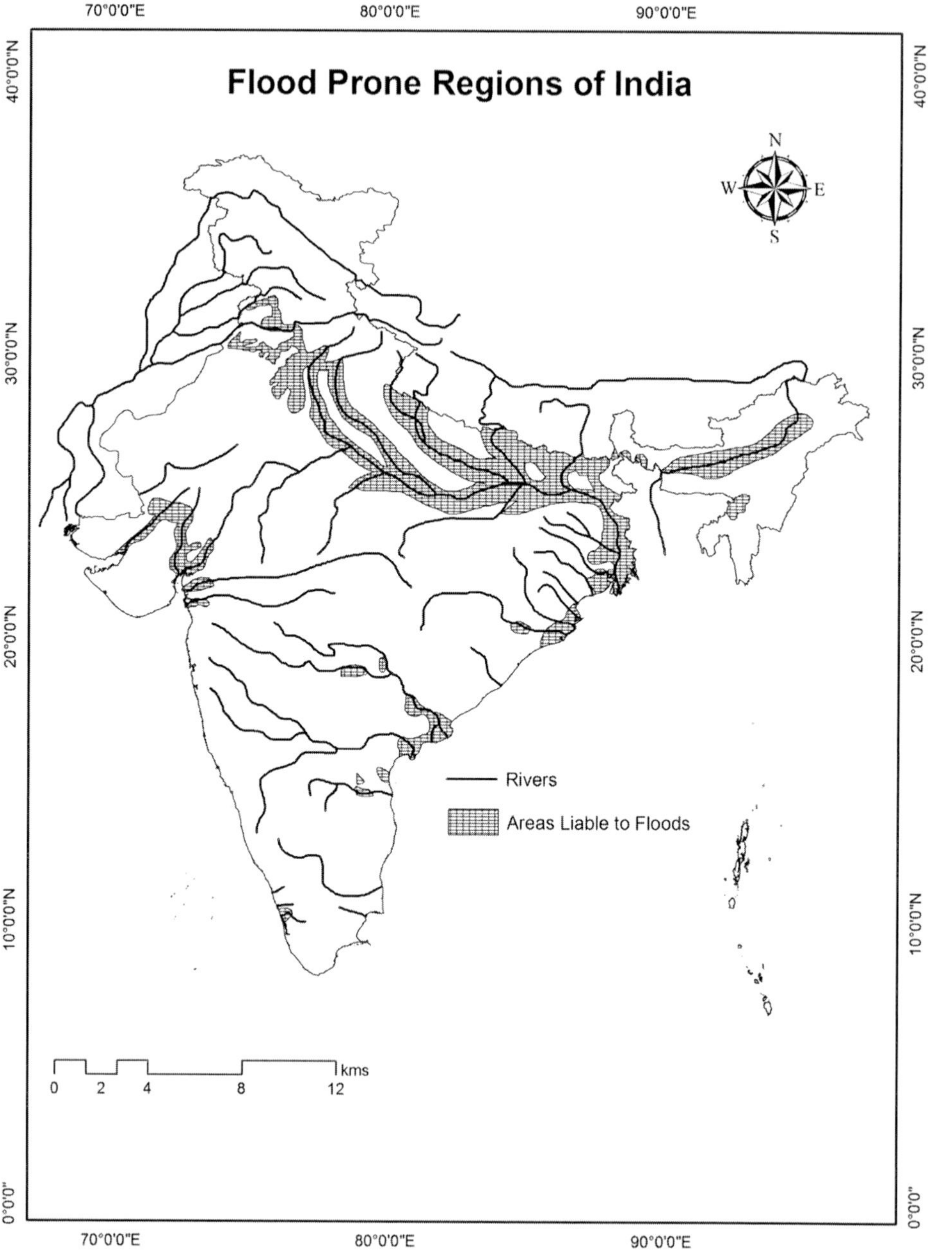

Figure 2.24 : Map showing Flood Zones of India

The major flood events of India during past 70 years is shown in Table 2.18. Similary past 50 years major flood events of the world is shown in Table 2.19. The deadly flood events of the world over many years is presented in Table 2.20.

Table 2.18: Major flood events of India

S.No.	Name of flood	Cause	Year	Damage
1	Kerala Floods	Heavy rainfall	2018	More than 36,000 people displaced across the state and 445 people died. In addition to all the 5 gates of the Idukki Dam were opened for the first time in history.
2	Brahmaputra Floods	Heavy rainfall	2017	Assam floods affected 1.8 million people in India as well as wildlife of Kaziranga National Park and Pobitora Wildlife Sanctuary
3	Kashmir Floods	Heavy rainfall	2014	Partially affected 1000 villages of Jammu and Kashmir state and nearly 200 people had died due to the floods and 350 villages had been submerged under water.
4	North Indian floods	Cloud burst	2013	Mainly Uttarakhand and nearby states. More than 5,700 people were presumed dead.
5	Uttarakhand Floods	Heavy rain	2013	The Kedarnath Temple was damaged and around 1000 people have died. It is one of the most disastrous floods in the history of India.
6	Himalayan Flash Floods	Cloud burst	3 Aug., 2012	At least 31 people dead and triggered landslides and flash floods.
7	Brahmaputra Floods	Heavy rain	2012	At least 124 people have been killed and 13 great India rhinos and around 500 animals have died in Kaziranga National Park.
8	Ladakh Floods	Cloud burst	6 Aug., 2010	Damaging 71 towns and villages in Leh and at least 255 people have died.
9	Bihar Flood	Heavy rain	2008	Affected over 2.3 million people.
10	Bihar flood	Heavy rain	2007	Bihar was afftected much more by this flood event in the last 30 years and had affected around 10 million people.
11	Gujarat Flood	Heavy rain	2005	The death toll was about 123 people state wide and more than 250,000 evacuated.

S.No.	Name of flood	Cause	Year	Damage
12	Chennai Floods	Heavy rain	2005	The worst floods to hit the city of Chennai, 50 people were killed due to flood and stampede.
13	Mumbai flood	Cloud burst	26 July 2005	At least 1,094 people dead and that day is still remembered as the day Mumbai came to a standstill. Loss of property loss was estimated at ₹ 550 crore.
14	Bihar Flood	Heavy rain	2004	Around 885 people and 3272 animals have been died and nearly 21 million people were affected in 20 districts of Bihar.
15	Bihar flood	Heavy rain	1987	It claimed lives of 1,399 humans, 302 animals and public property worth INR ₹ 68 billion.
16	Gujarat flood	Heavy rain	1979	Due to over flow of Machchhu River burst, tand had flooded the Rajkot district of Gujarat. Exact figure of loss of lives is unknown, but it is estimated between 1800 and 2500 people died.
17	Madras flood	Heavy rain	1943	Overflowed Coovum and the Adyar rivers was observed. Damage caused to life and property was immense however estimate figure is unknown.

Table 2.19 : Major flood events of the world

S.No.	Year	Place	Damage
1	August 1975	Henan Province, China	26,000 people died
2	November 13, 1985	Town of Armero, Colombia	23,000 people died
3	December 15, 1999	Estado Vargas, Venezuela	>5,000 people died
4	1953	North Sea flood of caused devastation in the United Kingdom, Belgium and especially the Netherlands	2,551 people died in U.K. and 1,835 people were died in Netherlands
5	August 29, 2005	New Orleans Flooding New Orleans, Louisiana.	1,464 people died
6	1963	Vajont Dam, flood, Italy	2,000 people died

S.No.	Year	Place	Damage
7	1972	Buffalo Creek Flood, United States	125 people were died
8	1889	Johnstown flood (dam break in Johnstown, Pennsylvania)	2,209 people died
9	May 1, 1966	Sgorigrad, Bulgaria	488 people died
10	1959	Malpasset dam collapse, France	421 people died
11	1936	Northeastern Akita, Japan	375 people died
12	1864	Great Sheffield Flood, England	270 people died
13	1923	Gleno Dam collapse, Italy	356 people died
14	1928	St. Francis Dam failure, United States	373–400 people died
15	August, 2007	North Korean floods	600 people died
16	May 14, 1868	Aichi, Japan	941 people died
17	2008	Santa Catarina floods Brazil.	135 people died
18	1985	Italy	268 people died
19	27 August, 1993	Qinghai, China	243 people died
20	June 9, 1972	Black Hills flood, United States	238 people died
21	July 2012	Floods in Krasnodar Krai, Russia	172 people died
22	2013	North Indian floods	More than 5,700 people were presumed dead.
23	2013	Uttarakhand Floods	The Kedarnath Temple was damaged and around 1000 people have died

Table 2.20 : Disastrous flood events of the world

S.No.	Place	Year	Event Name	Death toll
1	China	1931	China floods	1,000,000–4,000,000
2	China	1887	Yellow River (Huang He) flood	900,000–2,000,000
3	China	1975	Failure of 62 dams	229,000
4	China	1935	Yangtze river flood	145,000
5	Netherlands	1530	St. Felix's Flood	more than 100,000
6	North Vietnam	1971	Hanoi and Red River Delta flood	100,000
7	China	1911	Yangtze River flood	100,000

S.No.	Place	Year	Event Name	Death toll
8	Netherlands	1287	St. Lucia's flood	50,000–80,000
9	Netherlands	1212	North Sea flood	60,000
10	Netherlands	1219	St. Marcellus flood	36,000

2.11 Drought

It is an abnormally dry period, when there is not enough water to support agriculturel human and animal life, urban or environmental water needs (Figure 2.25). Drought is the consequence of a natural reduction in the amount of precipitation over an extended period of time. It is is a natural phenomenon and simply defined as "prolonged dry weather".

Around 68 per cent arable land of the country is the prone to drought of varying degrees. Drought prone areas comprise 108.11 million hectares out of a total land area of 329 Million hectares. About 50 million people are affected annually by drought. Of approximately 90 million hectares of rain-fed areas, about 40 million hectares are prone to scanty or no rain. It typically strikes arid areas of Rajasthan and Gujarat states. Drought is not uncommon in certain districts of Uttar Pradesh, Madhya Pradesh, Odisha and Andhra Pradesh.

Figure 2.25 : Drought affected area in India
[*Source:* Online search free download image]

Drought is not a purely physical phenomenon, but instead is an interplay between natural water availability and human demands for water supply.

Classification of Drought

- **Based on the degree of water stress**

  a) Severe b) Moderate c) Mild

- **Based on the duration of water stress**

  a) Permanent b) Seasonal c) Contingent
  d) Invisible

- **Based on the timing of water stress**

  a) Early season b) Mid season c) Terminal

- **Based on relevance to the users**

  a) Atmospheric b) Meteorological c) Hydrological

  d) Agricultural e) Physiological

i) Meteorological drought

Meteorological drought occurred, when the seasonal rainfall over an area is less than 75 per cent of its long-term average (LPA). It is caused, when there is a prolonged period with below average precipitation. It is further classified into different categories based on the percentage rainfall departure from its LPA.

Based on seasonal rainfall departure from Normal (LPA)

Mild	> 25 percent
Moderate	> 25–50 percent
Severe	> 50 percent
Disastrous	> 60 percent

ii) Hydrologic drought

It is drought situation, when there is marked depletion of surface water (pond, river etc) and consequent drying up water reservoirs, such as aquifers, lakes, and water level in the reservoirs falls below the statistical average. This condition can arise, when the meteorological drought prolonged over an extended period.

iii) Agricultural drought

It is a drought situation when there is insufficient soil moisture for fulfillment of crop water demand to average crop production. This condition can arise due to low soil moisture and types of soil characteristics.

Agricultural drought is further classified into different types and is given below:

Early season drought: It is caused due to delay onset of monsoon in an area/zone, which affects time of sowing of crops and crop durations.

Mid-season drought: It is caused due to breaks in monsoon rainfall over an area/zone and occurred during vegetative stage of the crop.

Late season or terminal drought: It is caused by early cessation of rainy season or early withdrawals of monsoon from an area/zone. This type of drought occurred during the reproductive phase of the crop.

Apparent drought: It is a situation, when there is a mismatch of the cropping pattern and rainfall distribution of an area/zone. The rainfall is adequate for growing of one crop, but not for other crops.

Invisible drought: It is a situation when in humid regions, rainfall is not match (below or less) with evapo-transpiration of the crops.

iv) Other types of drought

Permanent drought: When the drought is a recurring phenomenon in an area/zone, than it is called permanent drought like arid region of India Rajasthan, Bundelkhand zone of Madhya Pradesh.

Socio-economic drought: It occurs, when physical water shortage starts to affect people, individually and collectively. It combines the impact of meteorological, hydrological and agricultural drought on society. Some examples of impact of drought are given below:

- 1900 India, killing between 250,000 and 3.25 million.
- 1921–22, Soviet Union, in which 250,000 to 5 million perished from starvation due to drought.
- 1928–30, Northwest China, resulting in over 3 million deaths by famine.
- 1936 and 1941, Sichuan Province, China, resulting in 5 million death, respectively.

2.11.1 Country's Vulnerability to Drought

Area (percent)	Annual Rainfall (mm)	Drought situation
33	< 750	Chronic
35	750–1125	Drought prone

Total 68 percent area of the country - prone to drought in varying degrees.

Rainfall Variability

a) *Spatial*

Area Rainfall amount

All India	117 cm
Mausineram	1140 cm
Churu	10 cm

b) ***Temporal***

June Pre-Monsoon(March–May)	10%
SouthWest Monsoon(January–September)	74%
NorthEast Monsoon(October–December)	13%
Winter Rains (January–February)	03%

2.11.2 Root (Important) cause of Agricultural Drought

1. Late onset of monsoon rains
2. Inadequate precipitation
3. Long dry spells
4. Erratic distribution of rainfall
5. Early withdrawal of rains

Anthropogenic causes: There are several anthropogenic causes like over grazing, deforestation and converting more area into desertification etc.

2.11.3 Drought Frequency and Distribution Pattern

The recurrence of drought in various parts of India is given in Table 2.21 .

Table 2.21 : Recurrence of drought in India

Recurrence of drought	Regions/state
Once in 4–5 years	Rajasthan, Punjab, Gujarat, Telengana and Bundelkhand region of Madhya Pradesh
Once 6–8 years	Haryana, Jammu and Kashmir, Uttar Pradesh (East) Sub-Himalayan West Bengal, Vidarbha (Interior), Karnataka (South) and Central Andhra Pradesh
Once in 10 years	Himachal Pradesh,Uttar Pradesh (West), Madhya Pradesh, Bihar (Plains), Chhattisgarh, Marathwada, Kokan, Rayalseema and Kerla
Once in 15 to 20 years	Madhya Maharashtra, South Assam, Tamilnadu (Interior), Karnataka (North) and Odisha
Very rare	Coastal Karnataka, Jharkhand Plateau and Bengal

Source : H.S. Mavi

- Around 68 per cent of India's total area is drought prone.
- 99 districts out of 615 districts are drought prone.
- 9 meteorological subdivisions out of 36 subdivisions are drought prone.
- 50 million people are annually affected by drought.

2.11.4 Drought Prone Area

Many parts of Indian subcontinent is drought prone and drought is a permanent feature in this zone (Figure 2.26).

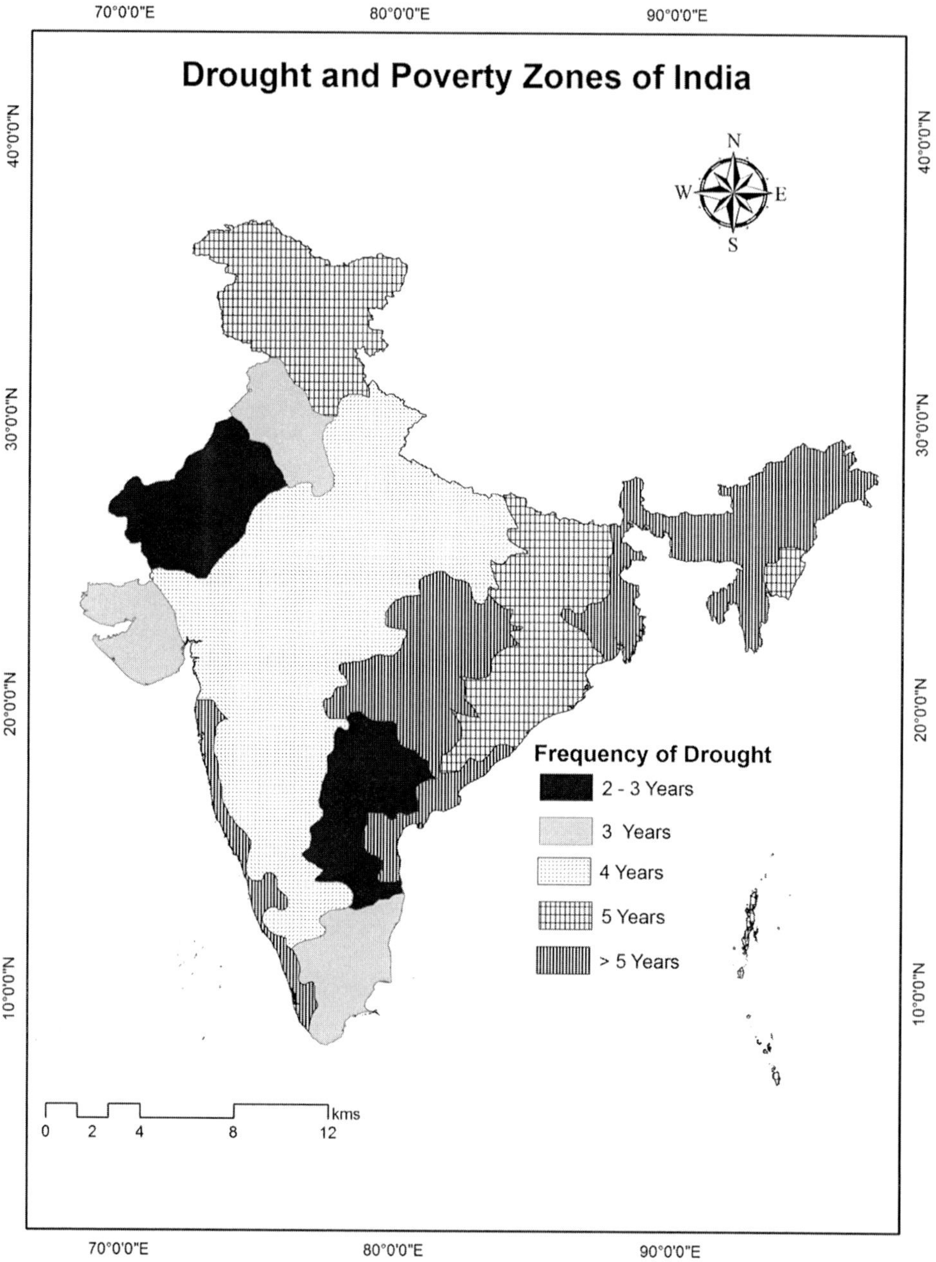

Figure 2.26 : Map showing drought prone states in India
(*Source* : MoES, IMD)

2.11.5 Adverse effects

The impacts slowly spread into social sphere due to shortage of drinking water, reduction in energy production, groundwater depletion, food shortage, health reduction and loss of life, increased poverty, and reduced quality of life. Several societal droughts may even lead to mass migration in search of food, fodder, water and work leading to famine, death and social unrest. The droughts have impacted the food grain production of India and shown in Figure 2.27.

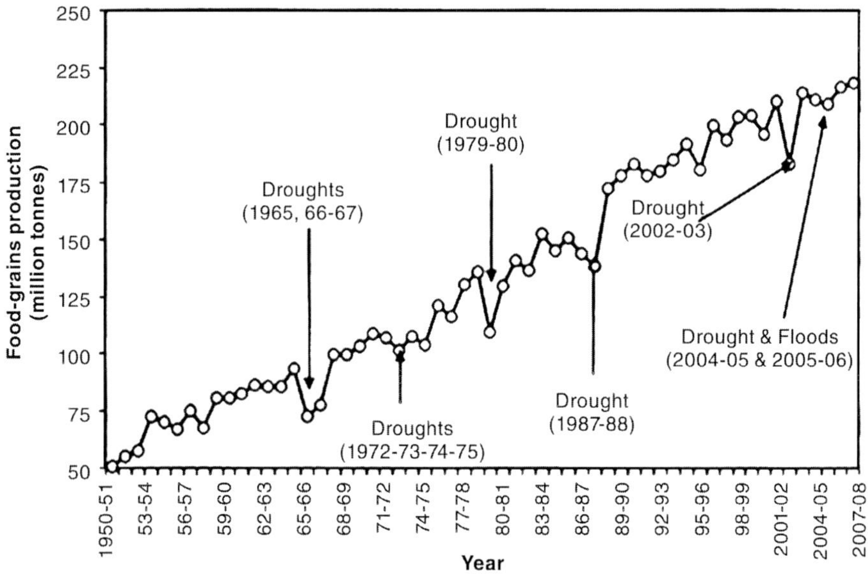

Figure 2.27 : Impact of droughts on Indian food grain production from 1950-51 to 2007-2008

2.11.6 Drought Declaration and Monitoring

The state government is responsible for drought declaration. Districts' Collectors can notify drought only after the state government declares drought in the state or part thereof. Generally, district collector based on two parameters rainfall deficiency and crop production estimates declared the drought in a district. The collector recommends for declaration of drought in the district only after the drought declaration of the state government.

As per of new manual of the department of Agriculture and Cooperation (DAC), Ministry of Agriculture (MA), Govt. of India (2009), drought declaration depends on combinations of factors. These factors are:

- Rainfall deficiency
- The extent of area sown
- Normalized Difference Vegetative Index (NDVI)
- Moisture adequacy index
- Extent of fodder supply
- Drinking water supply
- Supply of foodgrains and prices situation of essential commodities.
- Movement of labour in search of employment
- Current wages of the area etc.

Satellite based optical and radar imagery are used widely in monitoring agriculture. National drought assessment and monitoring system (NADAMS) : Mahalanobis National Crop Forecasting Centre (MNCFC) used NDVI, SWIR based indices for crop monitoring and SPI, standardized runoff index (SRI), for drought. IMD defines meteorological drought based on rainfall deficiency (SWM period) on sub-division wise basis. It monitors the incidence, spread, intensification and cessation of drought on a weekly time scale based on aridity anomaly index (AAI) and standardized precipitation index (SPI) on monthly basis.

2.11.7 Drought Events in India

India has a long history of droughts. There were 26 major droughts from 1800 to 2020, when the All India Summer Monsoon Rainfall (AISMR) was found to be lesser than the mean rainfall for the country. Even the most recent drought was disastrous for India, affecting agriculture and water resources across the country. Lists of drought years in India are shown in table 2.22.

Table 2.22 : Drought events in India

Period	Drought years	Number of drought
1801–1830	1801, 1804, 1806, 1812, 1819, 1825	6
1831–1860	1832, 1833, 1837, 1853, 1860	5
1861–1890	1862, 1866, 1868, 1873, 1877, 1883	6
1891–1920	1891, 1897, 1899, 1901, 1904, 1905, 1907, 1911, 1918, 1920	10
1921–1950	1939, 1941	2
1951–1980	1951, 1965, 1966, 1971, 1972, 1974, 1979	7
1981–2010	1982, 1987, 2002, 2009	4
2011–2020	2012,2013,2015	3

In India, during the time span of 1870 to 2020, eighteen meteorological and sixteen hydrological droughts occurred. The most severe meteorological droughts were in the years 1876, 1899, 1918, 1965 and 2000, while the five worst hydrological droughts occurred in the years 1876, 1899, 1918, 1965 and 2000. The drought of 1899 can be classified as meteorological as well as hydrological, and was the most severe documented drought India has ever experienced to date. Lists of severe droughts year in India is given in Table 2.23.

Table 2.23: Severe drought years in India

Year (Area percent)	affected in	(percent departure of rainfall in affected area
1877	60	-41
1899	68	-45
1918	71	-49

Year	(Area affected in percent)	(percent departure of rainfall in affected area
1965	41	-36
1972	47	-35
1979	45	-38
1987	50	-45
2009	47	-41
2012		

Questions for chapter evaluation

Descriptive questions

1. What are the differences between natural and manmade disaster?
2. Give examples of manmade disaster.
3. What do you mean by a cyclone?
4. Define Earthquake.
5. Define earthquake fault zone.
6. What do mean by ground shaking?
7. What are the effects of earthquake?
8. Write a note on earthquake measurement scales.
9. Write short notes on landslides and liquefaction.
10. Define Tsunamis.
11. Write about Tsunamis mitigation measures.
12. Define drought.
13. Write a note on drought management.
14. Discussed about the drought frequency in India.
15. How drought is declared?
16. Define flood and write about flood prone-area of India.
17. Write about the type of flood.
18. What is an avalanche event and names the most affected Indian states.
19. What is cloud burst?
20. List some cloud burst events.
21. How cyclone naming has been done?
22. What do you understand by super cyclone?
23. List most devastating cyclonic events.
24. Discussed the major hazards that were occurred in your state.
25. List the recent natural hazard of your state/district.

Objective questions

Fill in the blanks

1. .... has divided India into five earthquake risk zone.
2. Ritcher scale measures the earthquake ....
3. Cyclone is a ....disaster.
4. The earthquake has a highest ....scale.
5. Cyclone Nilam occurred in November ....
6. Drought in Bundelkhand region of Madhya Pradesh occurred in every .... years.
7. Around .... per cent area of India is flood-prone.
8. The recent Tsunami occurred in India in the year.......
9. Avalanche are common in Himalayan region above .... meter.
10. .... is an important factor to cause landslide in India.
11. Jal is a name of ....
12. Odisha supper cyclone occurred in the year ....
13. Hazard is .... phenomena or event.

Mention true and false against each of the questions given below:

a. Mantle is the harder part of earth structure.
b. The intensity scale measures the effect of an earthquake on the objects.
c. The highest magnitude of earthquake occurred in Bihar.
d. Tsunami warning is possible in India.
e. The deadly cyclone that was hit the Andhra coast in the year 2015.
f. Cloud burst is a manmade disaster.
g. The hilly region of Himalayas is moderately affected by landslides in India.
h. Avalanches are the biggest disaster in hilly region of India.
i. Flash flood is associated with scattered rainfall in a region.
j. Bundelkhand zone of Madhya Pradesh is a region of permanent drought.
k. Bihar is a drought-prone state in India.

CHAPTER **3**

EXTREME WEATHER EVENTS AND CLIMATE CHANGE

3.0 Extreme Weather

Extreme weather is a weather event which that has significantly different from the average or usual weather pattern. It may take place over one day or a period of time. A cold and heatwave is two examples of extreme weather in North India.

Many of the extreme weather events, *i.e.,* cold or heat waves are not visible as other forms of severe weather like cyclone, thunderstorms and heavy snowfall. Extreme events such as heatwaves, droughts, cyclone, and snowfall have affected India.

Some types of extreme weather events are listed below:

- Flood: heavy rainfall
- Cyclone, snowfall, hail.
- Strong wind, *i.e.,* squall, Gale.
- Blizzard: heavy snow, ice, cold temperatures.
- Dust storm: strong winds, arid conditions.
- Flood: heavy rainfall.
- Hail storm: cold or warm temperatures, rain, ice.
- Dust storm: Hot weather, heatwave
- Ice storm: freezing rain.
- Heat and cold waves

3.1 Extreme Weather Events

Thus extreme weather event is unexpected, unusual, severe or non-normal weather events. The major events of extreme weather in India are shown in Table 3.1.

Table 3.1. Top weather related hazards in India and their occurrences (1978–2006)

S.No.	Event	Total No.	Mortality
1	Flood	2973	43573
2	Cyclone	62	18853

S.No.	Event	Total No.	Mortality
3	Heatwave	373	8514
4	Cold wave	376	6668
5	Thunderstorm	1123	3490
6	Lightning	1353	3395
7	Hail storm	934	989
8	Dust storm	146	584
9	Squall	284	362
10	Gale	245	291

[*Source:* Singh and Patwardhan, APCBEE Procedia 1 (2012) 258-262]

3.1.1 Heatwaves

A heat wave is a prolonged period of high temperature and dry winds, which may be accompanied by low humidity. It is not a visible form of severe weather.

It is defined as appreciable rise in maximum temperature and the criteria for describing heat wave as follow:

Criteria for heatwave: Heatwave need not be considered till maximum temperature of a station reaches at least 40°C for Plains and at least 30°C for hilly regions.

Classification of heat wave: Based upon station normal maximum temperature, criteria are defined for different types of heatwaves and are described below:

When normal maximum temperature of a station/area is less than or equal to 40°C

Nomenclature	**Past 24 hours change** (Departure from normal)
Normal	1°C to -1°C
Moderate heatwave	+5°C to + 6°C
Severe heatwave	+7°C or less

a) When normal maximum temperature of a station is more than 40°C

Nomenclature	**Past 24 hours change** (Departure from normal)
Normal	1°C to - 1°C
Heatwave	+3°C to + 4°C
Severe heatwave	+5°C or more

b) When actual maximum temperature remains 45°C or more for two days irrespective of normal maximum temperature, heatwave should be declared.

The major heat events in India and the world are shown in table 3.2 and 3.3, respectively.

3.1.2 Hot Day

In the northern plains of the country, dust in suspension occurs in many years for several days, bringing minimum temperature much higher than normal and keeping the maximum temperature around or slightly above normal. Sometimes increase in humidity also adds to this discomfort. Nights do not get cooled and become uncomfortable. To cover this situation, hot day concept has been introduced as given below:

Whenever the maximum temperature remains 40°C or more and minimum temperature is 5°C or more above normal, it may be defined as hot day, provided it is not satisfying the heatwave criteria given above.

Criteria for describing hot day for coastal stations

When maximum temperature departure is 5°C or more from normal, Hot day may be described irrespective of the threshold value of 40°C. If the threshold value of 40°C is reached, Heat Wave may be declared.

When a station satisfies both the Heatwave and Hot Day criteria, then Heatwave should be given higher priority and be declared.

3.1.3 Hot Wind

Hot wind reduces moisture causing dehydration and a prolonged exposure may prove to be fatal.

The phenomena of Loo over the plains of northwest India is very well known. It is also described in the weather bulletins and appropriate warnings are being issued.

Table 3.2 : Heatwave events in India

S.No.	Year	Maximum temperature (°C)	Death toll
1	1956	50.6	
2	1995	45.5	>350
3	1998	49.5	2,500
4	2002	49.0	1,030
5	2015	49.4	2,500
6	2016	51.0	1,600
7	2019	50.8	184

Table 3.3: Major heatwave events of world

Heatwave Event	Location	Year	Death toll
European heatwave	Europe	2003	70,000
Russian heatwave	Russia	2010	56,000
United States heatwave	United States	1988	5,000–10,000
United States heatwave	United States	1980	1,700–5,000

Heatwave Event	Location	Year	Death toll
Japanese heatwave	Japan	2010	1,718
Southern India heatwave	India	2003	1,500
Los Angeles heatwave	United States	1955	946
New York City heatwave	United States	1972	891
Chicago heatwave	United States	1995	739
Argentina heatwave	Argentina	1900	475
South East Asia	Taiwan	2016	85

3.1.4 Cold Wave

Cold waves are recurrent phenomenon in North India. Thousands of people die of cold and related diseases every year, most of them are poor of urban areas in northern parts of the country.

Criteria for Cold wave: The cold wave is defined as appreciable fall in minimum temperature and the criteria for describing cold wave is as follow:

(a) **When the normal minimum temperature of a station/area is 10°C or more**

Nomenclature Past 24 hours change (Departure from normal)

Normal	1°C to -1°C
Below normal	-2°C
Appreciable below normal	-3°C to -4°C
Moderate cold wave	-5°C to -6°C
Severe cold wave	-7°C or less

(b) **When the normal minimum temperature of a station/area is more than 10°C**

Nomenclature Past 24 hours change (Departure from normal)

Normal	1°C to -1°C
Below normal	-2°C
Cold wave	-3°C to - 4°C
Severe cold wave	-5°C or more

Rapid fall of temperature during 24 hour period and it is determined by the rate at which the temperature falls. The loss of life due to heat and cold waves in India is shown in Table 3.4.

Table 3.4: Human death toll in India due to heat and cold wave conditions

S.No.	Year	Human deaths due to heatwaves	Human deaths due to cold waves
1	1980	156	185
2	1981	33	192
3	1982	16	63

S.No.	Year	Human deaths due to heatwaves	Human deaths due to cold waves
4	1983	187	488
5	1984	58	155
6	1985	142	494
7	1986	156	276
8	1987	87	105
9	1989	43	215
10	1990	2	82
11	1992	114	303
12	1993	30	63
13	1996	20	68
14	1997	21	140
15	1999	119	222
16	2000	55	368
17	2001	56	490
18	2004	117	462
19	2008	111	114
20	2009	216	79
21	2010	242	450
22	2011	12	722
23	2018	16	136

[*Source* : IMD, New Delhi]

Cold weather injury

Long periods of extreme cold weather combined with other meteorological phenomena are detrimental to many sectors of the economy, especially agricultural and can result in the loss of winter crops, fruit crops and vegetables due to frost injury.

Studies of plant injury caused by extremely cold weather reported that under low temperature, a plant basically dries out and the protoplasm dies. This happens because the concentration of cell sap rises, the distance between macromolecules is altered and the pro-accumulate within the plant's tissues and it is this that causes the protoplasm to die. The degree of damage depends on the intensity and duration of dangerous frosts as well as the stage of the plant's development.

3.2 Climate

Average meteorological conditions over a specified time period, usually at least a month, resulting from interactions among the atmosphere, oceans, and land surface. Climate variations occur over a wide range of spatial and temporal scales.

3.2.1 Climate Change

A change of climate, which is attributed directly or indirectly to human activity that alters the composition of the global atmosphere and which is in addition to natural climate variability observed over comparable time periods.

Changes in regional, global climate characteristics, including temperature, humidity, rainfall, wind, and any severe weather events is known as climate change. This change is the change related to either change in the absolute value (Normal) or in its standard deviation of the weather parameters.

Climate change refers to climatic conditions of an area. Climate change is related to warm or cool temperatures, whereas "global warming" refers to increasing global temperatures.

3.2.2 Climate Extremes

Used to represent weather extremes, but viewed over seasons (*e.g.*, droughts), or longer periods.

3.2.3 Climate Variability

It refers to variations or deviations from the mean state of the climate or temporal variations of the atmosphere-ocean system around a mean state measure over a long period of time. The term "natural climate variability" is further used to identify climate variations that are not atributed to or influenced by any activity related to humans. However, it is recognized that such "internal or natural variability" could be affected by external factors driving climate change such as changes in the atmospheric concentration of greenhouse gases. The El Niño-Southern Oscillation (ENSO) phenomenon is a good example of the variability in the coupled oceanic and atmosphere system that is a central factor in short-term climate variability and the interannual timescale.

3.3 Global Warming

Overall warming of the planet or its nearby atmosphere based on average temperature is know as global warming. This warming leads to a variety of negative effects, including:

1. Melting of glaciers and mountain snowcaps that feed the world's rivers and supply a large portion of the freshwater used for drinking and irrigation.
2. A rise in sea levels due to the melting of the land-based ice sheets in Greenland and Antarctica,.
3. Increasingly "extreme weather" events, such as heatwaves, droughts, floods, and severe storms.
4. Lowered agricultural productivity due to less favourable weather conditions, less available irrigation water, increased heat stress to plants, and an increase in pest activity due to warmer temperatures.
5. Increases in vector-borne infectious diseases like malaria and Lyme Disease.

6. Large numbers of extinctions of higher-level species due to their inability to adapt to rapidly changing climate and habitat conditions.

3.4 Greenhouse Gases

Many chemical compounds found in the earth's atmosphere and acts as "greenhouse gases." These gases allow sunlight to enter (short wave radiation) the atmosphere freely, but when this sunlight strikes the earth's surface, some of it is reflected towards space as infrared radiation (heat). Greenhouse gases absorb this infrared radiation and trap the heat in the atmosphere. The effect is known as greenhouse effect and many gases exhibit these "greenhouse" properties. Some of them occur in nature (water vapour, carbon dioxide, methane, and nitrous oxide), while others are exclusively generated and added in the atmosphere by many human activities (like CO_2, N_2O etc.).

3.4.1 Sources of Greenhouse Gases

The greenhouses are generated by industrial growth, fuel used for electricity generation, and use of petroleum products and burning of fossil fuel. Energy-related carbon dioxide emissions, resulting from petroleum and natural gas, represent 82 percent of total U.S. human-made greenhouse gas emissions (Figure 3.1).

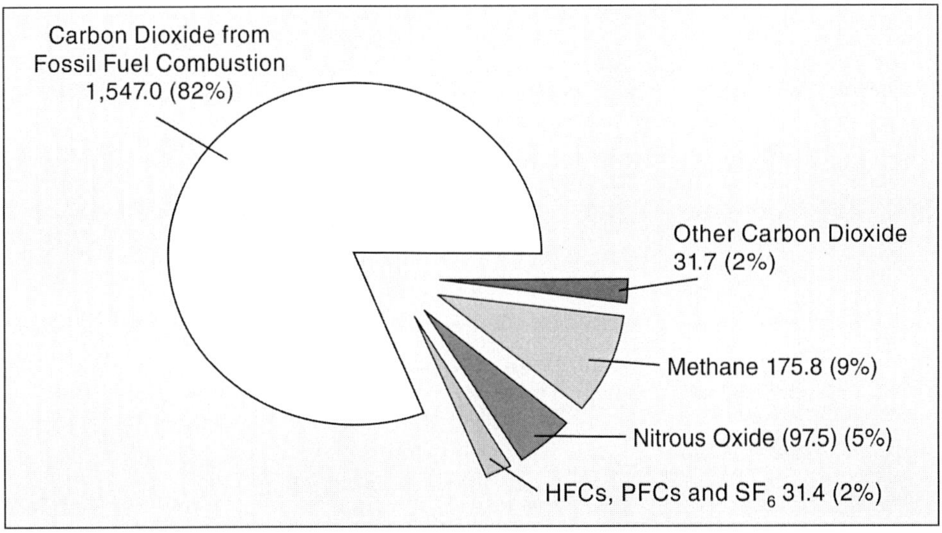

Figure 3.1 : Greenhouse gases and their contribution to global warming and climate change [*Source:* Energy Information Administration Emissions of Greenhouse Gases in the United States 2001 (Washington, DC, 2002)]

Other greenhouse gas, like methane, comes from landfills, coal mines, oil and gas operations, and agriculture; it represents 9 percent of total emissions. Nitrous oxide (5 percent of total emissions) is emitted from burning fossil fuels and through the use of certain fertilizers and industrial processes. Human-made gases (2 percent of total emissions) are released as byproducts of industrial processes.

Levels of several important greenhouse gases have increased by about 25 percent since large-scale industrialization began around 150 years ago (Figure 3.2). During the past 20 years, about three-quarters of human-made carbon dioxide emissions were from fossil fuels burning.

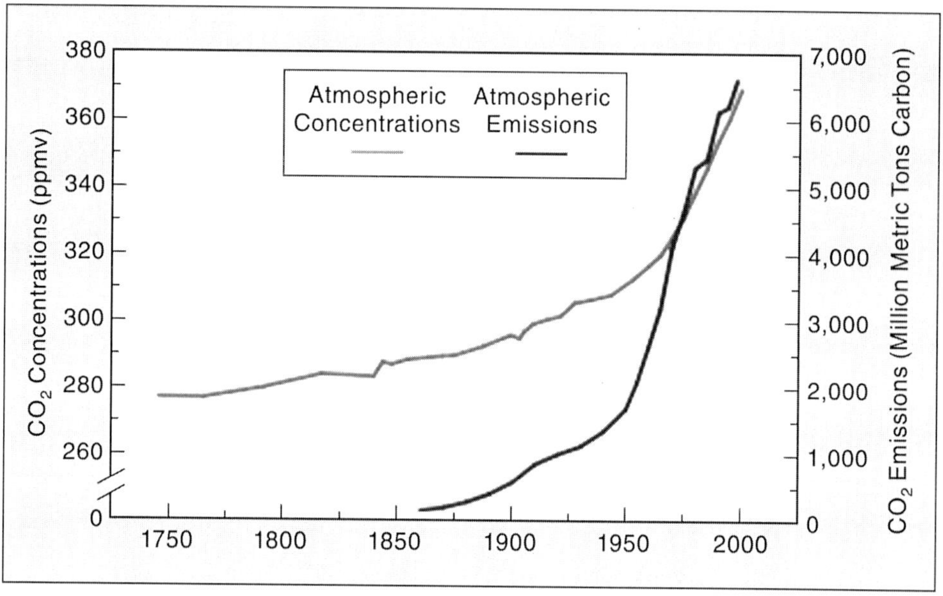

Figure 3.2 : Trends in Atmospheric Concentrations and Anthropogenic Emissions of Carbon Dioxide [*Source:* Oak Ridge National Laboratory. Carbon Dioxide Information Analysis Centre, http://cdrac.esd.oml.gov/.]

3.4.2 The Greenhouse Effect

The greenhouse effect is caused by gases in the atmosphere, which has the ability to absorb the sun's energy that is usually radiated back into space from Earth. Energy from the sun comes into the earth as short-wave radiation; some are absorbed and some are radiated back as long-wave radiation. The 'greenhouse gases' allow the short-wave radiation to pass through to Earth, but absorb the long-wave radiation that is reflected to space. These gases include naturally occurring gases, primarily water vapour, carbon dioxide, methane, nitrous oxides as well as industrial chemicals such as chlorofluorocarbons (CFCs).

Human activities have increased the atmospheric concentration of these gases well beyond their natural levels, and have introduced new greenhouse gases, such as CFCs, HFCs, SF_4 etc. One of the major greenhouse gases from human sources is carbon dioxide (CO_2). While CO_2 is naturally occurring, its concentration in the atmosphere is rapidly increasing because of the burning of fossil fuels– oil, coal and gas.

3.4.3 The Glass House Effect

The greenhouse effect can be visualized as glass house effect also. The glass/green housesroofs are generally made up by giant white glass. The heat of the sun penetrates through the glass. Some of the heat is absorbed by the glass, and some of it is radiated back towards space. The radiated heat reaches the glass roof and is prevented from dispersing any further and thus trap the heat and increased the temperature inside the glasshouse.

Similarly, the earth is surrounded by a blanket of gases. This blanket traps energy in the atmosphere, much the same way as glass traps heat inside a greenhouse. This results in an accumulation of energy, and the overall warming of the atmosphere. The 'greenhouse effect' is the popular expression for the above process.

3.5 Impacts of Climate Change

Climate models show that atmospheric temperatures will increase by between 1.5 and 4.5°C by the year 2100 if a "wait and see and do nothing" approach is adopted. The impact of climate change as indicated by different models is that there will be a drastic temperature rise both in terms of its intensity and its rate of increase. It is projected to have drastic social, economic and ecological implications:

- Because of thermal expansion of the water and melting of continental glaciers sea levels would rise, possibly as much as two feet, by the end of next century.
- Rising temperatures could lead to changes in regional wind systems which would influence global rainfall distribution and lead to the redistribution and frequency of floods, droughts and forest fires. Windstorms and hurricanes could become more frequent and more intense.
- Increased sea temperatures would cause coral bleaching and the destruction of coral reefs around the world.
- Climate change would create favourable conditions for growth in insect populations. This would likely harm agriculture and human health, and result in a spread of malaria and other tropical diseases.
- Water supplies would become disrupted in some regions, particularly in already vulnerable, arid areas.

3.5.1 Consequences of Potential Global Climate Changes on Agricultural Production

Effect of temperature on crop yield

Climate change is likely to affect crops differently from region to region. For example, average crop yield is expected to drop down by 20 to 30 percent in India and upto 50 per cent in Pakistan according to the UKMO scenario, whereas corn production in Europe is expected to grow up to 25 percent in optimum hydrologic conditions. The effects are also different depending on the latitude. In temperate countries, effects are found less negative or even rather beneficial, while in tropical and desertic countries they tend to be adverse.

Effect on growing period

An increase in temperature will speed up development. In the case of an annual crop, the duration between sowing and harvesting will shorten (for example, the duration in order to harvest wheat and mustard could shorten). The shortening of such a cycle would have an adverse effect on productivity because senescence would occur sooner.

Effect of atmospheric carbon dioxide on yield

Rising carbon dioxide (CO_2) concentration in the atmosphere can have both positive and negative consequences. Increased CO_2 is expected to have positive physiological effects by increasing the rate of photosynthesis. Currently, the amount of carbon dioxide in the atmosphere is 400 parts per million. The effects of increase in carbon dioxide would be higher on C_3 crops (such as wheat) than on C_4 crops (such as maize) because the former is more susceptible to carbon dioxide shortage. Under optimum conditions of temperature and humidity, the yield increase could more if the levels of carbon dioxide are doubled. However, other studies also show a change in harvest quality. The growth improvement in C_3 plants could favour vegetative biomass on grain biomass; thus leading to a decrease in grain yield.

A higher ambient level of carbon dioxide allows plants to respire CO_2 more readily with a less stomatal opening. Plants lose water due to transpiration through open stomata, so they are adapted to conserve water by opening the stomata only as necessary.

Higher carbon dioxide levels within cells can allow the stomata to be closed longer without suffering photorespiration, an effect of too much oxygen in ratio to carbon dioxide in the plant cell's chloroplasts.

Effect on quality

According to the IPCC's reports, climate change impacts on grain and forage quality. For rice, the amylose content of the grain (cooking quality) is increased under elevated CO_2. Cooked rice grain from plants grown in high CO_2 environments would be firmer than that from today's plants. However, concentrations of iron and zinc, which are important for human nutrition, would be lower. Moreover, the protein content of the grain decreases under combined increases of temperature and CO_2.

Studies have shown that higher CO_2 levels lead to reduced plant uptake of nitrogen resulting in crops with lower nutritional value. This would primarily impact on populations in poorer countries less able to compensate by eating more food, more varied diets, or possibly taking supplements.

Reduced nitrogen content in grazing plants has also been shown to reduce animal productivity in sheep, which depend on microbes in their gut to digest plants, which in turn depend on nitrogen intake.

Effect on agricultural land availability

Sea levels are expected to get up to one meter higher by 2100. A rise in sea level would result in agricultural land loss, in particular areas such as South East Asia. Erosion, submergence of shorelines, salinity of the water due to the increased sea levels, could mainly affect agriculture through inundation of low-lying lands. However, increase in area of arable land near the poles may be possible due to the reduction of the amount of frozen lands.

Effect on erosion and fertility

Due to global warming, soil degradation will be more and also would affect soil fertility. A doubling of carbon is likely to imply higher storage of nitrogen in soils as nitrates, thus providing higher fertilizing elements for plants, providing better yields.

Due to the increase in precipitations and in the intensity of the rain would probably result in greater risks to soil erosion. The increase in the temperature would induce a greater rate in the production of minerals, lessening the soil organic matter content, the atmospheric CO_2 concentration would tend to increase it.

Efects on pests, diseases and weeds

Increase of atmospheric humidity and higher temperatures would favour the development of fungal diseases. Similarly, because of higher temperatures and humidity, there could be increased pressure from insects and disease vectors.

Weeds would be benefited from carbonaceous fertilization. It would be possible that weed killers could be more effective with the temperature increase.

3.6 Ozone and UV-B

Depletion/destruction of the ozone layer by human-made chemicals, such as CFCs, is allowing increased levels of harmful UV-B radiation to reach the surface of the earth. Increased levels of UV-B radiation could reduce the density of plankton in the oceans. Since plankton are the primary carbon sink of the planet, reduction in their density could result in less CO_2 being absorbed from the atmosphere.

Scientists reported that agriculture could be affected by any decrease in stratospheric ozone, which could increase biologically dangerous ultraviolet radiation B. Excess ultraviolet radiation B can directly affect plant physiology and cause massive amounts of mutations, and indirectly through changed pollinator behaviour. In addition, a possible effect of rising temperatures is significantly higher levels of ground-level ozone, which would substantially lower the crop yields.

3.7 Climate Change Mitigation

The big challenge for scientific community is how to mitigate the effects of climate change. A number of strategies have been proposed to deal with this problem. This may be achieved by cutting the CO_2 emissions by switching from burning fossil fuels for using non-fossil-fuel sources of energy, such as nuclear energy, wave and wind power, and geothermal sources. To reduce the concentration of CO_2 from atmosphere is either draw-down the amount of CO_2 from the atmosphere or injecting CO_2 deep in the ocean or into geological strata. Another way is by increasing cloud-top albedo by sulphate injection or seawater seeding is designed to reflect incoming solar radiation and hence reduce global warming. In addition to these geo-engineering approaches, there is a suite of biology-based potential solutions (bio-geo-engineering) to mitigating global environment change. There are other options also and given below:

- Reducing CO_2 emission through Soil Carbon Sequestration
- Reducing atmospheric CO_2 concentrations by Afforestation
- Reducing surface warming by increasing crop albedo

3.8 Resilience of Agriculture Sector to Climate Change Through Realize, Reorganize and Adaption

Organic Agriculture helps Adaptation to Climate Change

Adaptation to climate change is a key issue for many developing countries. Farmers will need to adapt their practices to deal with changing temperatures and more frequent extreme weather events. Organic agriculture and green farming will improve soil quality and efficient water use also strengthen agroecosystems, which enables them to better respond to change.

3.9 Extreme Events in Bundelkhand Agroclimatic Zone of Madhya Pradesh

Extreme weather events in general; are determinant for agricultural production and in particular in *rainfed* agro-ecosystem. It has been reported that extreme weather events are likely to be increased in central India, which affects the crop yield. To have an idea the frequency of the extreme weather events in the three districts of Bundelkhand Agroclimatic zone of Madhya Pradesh, (Figure 3.3) long term weather data was analyzed and presented below:

Figure 3.3 : Bundelkhand Agroclimatic zone of Madhya Pradesh, India

Daily temperatures (maximum and minimum) and rainfall data during the period 1971–2010 of the three districts were collected and screened for extreme weather events on a decadal scale. The district-level extreme weather events days were calculated on decadal basis. The following criteria were adopted:

Hot day : when the daily maximum temperature >45°C,
Cold day : when the daily minimum temperature <2°C,
Heavy rainy day : when the daily rainfall > 50mm,
Rainy day : when the daily rainfall ≥ 2.5mm

Table 3.5: Decadal variability extreme weather events at Tikamgarh District

Decade	Number of hot days	Number of frost days	Number of heavy rainy days	Number of rainy days
1971–80	33	13	47	484
1981–90	53	03	49	430
1991–2000	65	04	48	497
2001–10	40	00	19	355

Table 3.6: Decadal variability extreme weather events at Chhatarpur District

Decade	Number of hot days	Number of frost days	Number of heavy rainy days	Number of rainy days
1971–80	66	07	58	495
1981–90	58	02	59	571
1991–2000	111	01	58	531
2001–10	112	02	51	481

Table 3.7: Decadal variability extreme weather events at Datia District

Decade	Number of hot days	Number of frost days	Number of heavy rainy days	Number of rainy days
1971–80	-	-	-	-
1981–90	43	00	43	458
1991–2000	78	01	37	466
2001–10	137	03	29	454

Source : Srivastava *et al.*, 2020, Mausam

The above reported data showed that heavy rainfall days were decreased sharply in Tikamgarh (from 47 to 19 days) and Datia (from 43 to 29 days) districts, while the decrease in rain was only observed in Tikamgarh. There was abrupt increase in number of hot days at Chhatarpur (from 66 to 112) and Datia (from 43 to 137) over the decades. The effects of increased number of hot days exhibit a larger impact on crop yield because of the increased maximum temperature. These effects are affecting the rate of phenology and reduced the grain-filling periods. Observations in controlled environment studies show that crop yield is greatly reduced by above normal temperatures during the grain-filling period.

Extreme weather at Tikamgarh district of Madhya Pradesh

The weekly weather extreme of temperature and rainfall were screened from the daily weather data from 1980 to 2018 of Tikamgarh district. The weekly extreme values with their occurrence dates are given below:

Standard Meteorological Week-1

Extreme weather Events	Temperature (°C)		Rain (mm)	Occurrence of the events		
	Maximum	Minimum		Date	Date	Date
Maximum value	30.6	14.7	31.0	03/01/1993	03/01/1993	02/01/1984
Minimum value	12.6	-0.6	0.0	01/01/2015	05/01/2011	—

Standard Meteorological Week-2

Extreme weather Events	Temperature (°C)		Rain (mm)	Occurrence of the events		
	Maximum	Minimum		Date	Date	Date
Maximum value	30.2	15.0	39.4	11/01/2008	13/01/2000	11/01/1987
Minimum value	15.2	-0.4	0.0	14/01/2014	08/01/2013	—

Standard Meteorological Week-3

Extreme weather Events	Temperature (°C)		Rain (mm)	Occurrence of the events		
	Maximum	Minimum		Date	Date	Date
Maximum value	32.4	16.5	22.8	17/01/2006	19/01/1980	17/01/1987
Minimum value	13.6	1.4	0.0	17/01/2014	21/01/2013	—

Standard Meteorological Week-4

Extreme weather Events	Temperature (°C)		Rain (mm)	Occurrence of the events		
	Maximum	Minimum		Date	Date	Date
Maximum value	33.7	31.0	54.0	28/01/2007	26/01/2009	22/01/1982
Minimum value	9.7	1.5	0.0	23/01/2009	22/01/2013	—

Standard Meteorological Week-5

Extreme weather Events	Temperature (°C)		Rain (mm)	Occurrence of the events		
	Maximum	Minimum		Date	Date	Date
Maximum value	33.6	17.5	9.2	30/01/2007	30/01/2007	02/02/2003
Minimum value	16.9	2.6	0.0	29/01/2015	30/01/2015	—

Standard Meteorological Week-6

Extreme weather Events	Temperature (°C)		Rain (mm)	Occurrence of the events		
	Maximum	Minimum		Date	Date	Date
Maximum value	32.5	17.2	41.0	08/02/1990	09/02/2010	05/02/2013
Minimum value	15.0	1.1	0.0	05/02/1999	10/02/2012	—

Standard Meteorological Week-7

Extreme weather Events	Temperature (°C)		Rain (mm)	Occurrence of the events		
	Maximum	Minimum		Date	Date	Date
Maximum value	37.1	18.2	46.6	14/02/1998	17/02/2001	16/02/2013
Minimum value	17.8	3.3	0.0	17/02/2013	14/02/2008	—

Standard Meteorological Week-8

Extreme weather Events	Temperature (°C)		Rain (mm)	Occurrence of the events		
	Maximum	Minimum		Date	Date	Date
Maximum value	37.2	19.6	31.0	25/02/2006	22/02/2005	19/02/2010
Minimum value	19.1	2.5	0.0	24/02/2014	24/02/1984	—

Standard Meteorological Week-9

Extreme weather Events	Temperature (°C)		Rain (mm)	Occurrence of the events		
	Maximum	Minimum		Date	Date	Date
Maximum value	38.2	19.5	22.0	04/03/2009	02/03/2001	27/02/2014
Minimum value	18.1	3.0	0.0	02/03/2015	27/02/1984	—

Standard Meteorological Week-10

Extreme weather Events	Temperature (°C)		Rain (mm)	Occurrence of the events		
	Maximum	Minimum		Date	Date	Date
Maximum value	37.7	24.0	28.0	08/03/1985	10/03/2001	09/03/2006
Minimum value	19.2	5.0	0.0	11/03/2006	05/03/1984	—

Standard Meteorological Week-11

Extreme weather Events	Temperature (°C)		Rain (mm)	Occurrence of the events		
	Maximum	Minimum		Date	Date	Date
Maximum value	40.0	21.2	23.0	17/03/2004	15/03/2003	15/03/2015
Minimum value	21.6	8.0	0.0	18/03/1998	14/03/1982	—

Standard Meteorological Week-12						
Extreme weather Events	Temperature (°C)		Rain (mm)	Occurrence of the events		
	Maximum	Minimum		Date	Date	Date
Maximum value	41.6	25.0	5.8	23/03/2010	22/03/1991	19/03/2017
Minimum value	24.3	8.4	0.0	21/03/1981	19/03/2017	—

Standard Meteorological Week-13						
Extreme weather Events	Temperature (°C)		Rain (mm)	Occurrence of the events		
	Maximum	Minimum		Date	Date	Date
Maximum value	41.5	24.7	23.0	01/04/1994	01/04/2010	27/03/1995
Minimum value	29.8	8.1	0.0	29/03/2017	28/03/1993	—

Standard Meteorological Week-14						
Extreme weather Events	Temperature (°C)		Rain (mm)	Occurrence of the events		
	Maximum	Minimum		Date	Date	Date
Maximum value	42.3	27.0	3.1	07/04/2002	06/04/2012	07/04/2015
Minimum value	30.2	7.6	0.0	05/04/2015	03/04/1990	—

Standard Meteorological Week-15						
Extreme weather Events	Temperature (°C)		Rain (mm)	Occurrence of the events		
	Maximum	Minimum		Date	Date	Date
Maximum value	47.1	26.0	15.6	10/04/1985	10/04/2010	12/04/1994
Minimum value	25.1	12.3	0.0	15/04/1983	10/04/2017	—

Standard Meteorological Week-16

Extreme weather Events	Temperature (°C)		Rain (mm)	Occurrence of the events		
	Maximum	Minimum		Date	Date	Date
Maximum value	46.4	29.5	11.2	20/04/1996	21/04/2010	16/04/1983
Minimum value	24.1	15.0	0.0	16/04/1983	21/04/1983	—

Standard Meteorological Week-17

Extreme weather Events	Temperature (°C)		Rain (mm)	Occurrence of the events		
	Maximum	Minimum		Date	Date	Date
Maximum value	45.3	31.2	0.7	29/04/2002	28/04/2010	23/04/2014
Minimum value	31.7	15.9	0.0	23/04/2013	24/04/2018	—

Standard Meteorological Week-18

Extreme weather Events	Temperature (°C)		Rain (mm)	Occurrence of the events		
	Maximum	Minimum		Date	Date	Date
Maximum value	47.0	30.0	1.6	03/05/1990	03/05/2002	01/05/2018
Minimum value	32.4	11.6	0.0	06/05/2001	02/05/1988	—

Standard Meteorological Week-19

Extreme weather Events	Temperature (°C)		Rain (mm)	Occurrence of the events		
	Maximum	Minimum		Date	Date	Date
Maximum value	46.0	32.3	4.4	12/05/1988	11/05/2002	10/05/1981
Minimum value	32.5	15.5	0.0	08/05/2000	07/05/1989	—

Standard Meteorological Week-20

Extreme weather Events	Temperature (°C)		Rain (mm)	Occurrence of the events		
	Maximum	Minimum		Date	Date	Date
Maximum value	46.9	32.4	20.8	19/05/2002	20/05/2002	18/05/2001
Minimum value	34.5	18.0	0.0	15/05/2015	20/05/2003	—

Standard Meteorological Week-21

Extreme weather Events	Temperature (°C)		Rain (mm)	Occurrence of the events		
	Maximum	Minimum		Date	Date	Date
Maximum value	48.8	34.4	24.0	27/05/1998	26/05/2010	23/05/2011
Minimum value	33.0	14.0	0.0	26/05/2005	25/05/2003	—

Standard Meteorological Week-22

Extreme weather Events	Temperature (°C)		Rain (mm)	Occurrence of the events		
	Maximum	Minimum		Date	Date	Date
Maximum value	48.7	34.6	42.0	27/05/1998	26/05/2010	23/05/2011
Minimum value	27.8	15.0	0.0	04/06/2011	29/05/2003	—

Standard Meteorological Week-23

Extreme weather Events	Temperature (°C)		Rain (mm)	Occurrence of the events		
	Maximum	Minimum		Date	Date	Date
Maximum value	48.3	34.8	68.0	05/06/1995	06/06/1995	08/06/1995
Minimum value	31.0	19.8	0.0	08/06/2000	09/06/1989	—

Standard Meteorological Week-24

Extreme weather Events	Temperature (°C)		Rain (mm)	Occurrence of the events		
	Maximum	Minimum		Date	Date	Date
Maximum value	46.4	33.1	61.0	13/06/1993	12/06/2007	16/06/2004
Minimum value	28.4	18.5	0.0	17/06/2013	16/06/1989	—

Standard Meteorological Week-25

Extreme weather Events	Temperature (°C)		Rain (mm)	Occurrence of the events		
	Maximum	Minimum		Date	Date	Date
Maximum value	45.7	33.2	25.5	20/06/1994	23/06/2010	19/06/2008
Minimum value	24.4	16.0	0.0	20/06/2008	23/06/1989	—

Standard Meteorological Week-26

Extreme weather Events	Temperature (°C)		Rain (mm)	Occurrence of the events		
	Maximum	Minimum		Date	Date	Date
Maximum value	45.1	30.5	156.8	27/06/2005	27/06/2010	29/06/2011
Minimum value	24.6	16.0	0.0	02/07/2013	27/06/2003	—

Standard Meteorological Week-27

Extreme weather Events	Temperature (°C)		Rain (mm)	Occurrence of the events		
	Maximum	Minimum		Date	Date	Date
Maximum value	44.2	30.2	121.6	08/071995	05/07/1982	03/07/1983
Minimum value	25.1	19.0	0.0	03/07/2013	09/07/2003	—

Standard Meteorological Week-28

Extreme weather Events	Temperature (°C)		Rain (mm)	Occurrence of the events		
	Maximum	Minimum		Date	Date	Date
Maximum value	43.5	31.4	162.8	11/07/2014	14/07/1995	14/07/1987
Minimum value	26.4	16.0	0.0	14/07/2017	13/07/1991	—

Standard Meteorological Week-29

Extreme weather Events	Temperature (°C)		Rain (mm)	Occurrence of the events		
	Maximum	Minimum		Date	Date	Date
Maximum value	40.2	31.2	115.4	17/07/2002	17/07/1995	20/07/2000
Minimum value	25.2	16.8	0.0	22/07/2011	19/07/1991	—

Standard Meteorological Week-30

Extreme weather Events	Temperature (°C)		Rain (mm)	Occurrence of the events		
	Maximum	Minimum		Date	Date	Date
Maximum value	40.0	34.4	155.0	30/07/2002	25/07/2005	27/07/2006
Minimum value	24.8	16.0	0.0	30/07/1999	27/07/1991	—

Standard Meteorological Week-31

Extreme weather Events	Temperature (°C)		Rain (mm)	Occurrence of the events		
	Maximum	Minimum		Date	Date	Date
Maximum value	40.0	28.4	370.0	01/08/2002	03/08/1987	02/08/1984
Minimum value	26.0	16.0	0.0	31/07/2012	01/08/1991	—

Standard Meteorological Week-32

Extreme weather Events	Temperature (°C)		Rain (mm)	Occurrence of the events		
	Maximum	Minimum		Date	Date	Date
Maximum value	37.4	28.0	225.0	09/08/1990	09/08/2009	07/08/1994
Minimum value	25.0	16.3	0.0	10/08/2013	09/08/1995	—

Standard Meteorological Week-33

Extreme weather Events	Temperature (°C)		Rain (mm)	Occurrence of the events		
	Maximum	Minimum		Date	Date	Date
Maximum value	37.8	27.5	165.8	20/08/1987	18/08/1987	14/08/1989
Minimum value	25.3	16.5	0.0	14/08/2012	17/08/1991	—

Standard Meteorological Week-34

Extreme weather Events	Temperature (°C)		Rain (mm)	Occurrence of the events		
	Maximum	Minimum		Date	Date	Date
Maximum value	41.0	27.2	105.1	23/08/2002	26/08/2001	20/08/2016
Minimum value	25.7	17.0	0.0	26/08/1982	26/08/1991	—

Standard Meteorological Week-35

Extreme weather Events	Temperature (°C)		Rain (mm)	Occurrence of the events		
	Maximum	Minimum		Date	Date	Date
Maximum value	27.5	27.6	143.6	30/08/1998	30/08/2001	02/09/1999
Minimum value	20.6	17.0	0.0	02/09/2002	28/08/1991	—

Standard Meteorological Week-36

Extreme weather Events	Temperature (°C)		Rain (mm)	Occurrence of the events		
	Maximum	Minimum		Date	Date	Date
Maximum value	37.2	29.6	121.0	09/09/2001	06/09/2008	09/09/1983
Minimum value	24.1	8.0	0.0	04/09/1982	10/09/2002	—

Standard Meteorological Week-37

Extreme weather Events	Temperature (°C)		Rain (mm)	Occurrence of the events		
	Maximum	Minimum		Date	Date	Date
Maximum value	39.3	26.3	170.6	15/09/1996	14/09/1984	12/09/1992
Minimum value	24.8	16.5	0.0	13/09/1992	15/09/1992	—

Standard Meteorological Week-38

Extreme weather Events	Temperature (°C)		Rain (mm)	Occurrence of the events		
	Maximum	Minimum		Date	Date	Date
Maximum value	39.4	26.3	248.2	22/09/2001	19/09/1996	18/09/1993
Minimum value	25.0	17.0	0.0	18/09/1990	23/09/1991	—

Standard Meteorological Week-39

Extreme weather Events	Temperature (°C)		Rain (mm)	Occurrence of the events		
	Maximum	Minimum		Date	Date	Date
Maximum value	39.0	26.1	93.2	25/09/2009	29/09/2009	28/09/1981
Minimum value	27.6	12.8	0.0	25/09/2007	30/09/1991	—

Standard Meteorological Week-40

Extreme weather Events	Temperature (°C)		Rain (mm)	Occurrence of the events		
	Maximum	Minimum		Date	Date	Date
Maximum value	38.3	25.6	108.0	05/10/1995	03/101988	06/10/2009
Minimum value	25.7	11.8	0.0	07/10/2009	04/10/1991	—

Standard Meteorological Week-41

Extreme weather Events	Temperature (°C)		Rain (mm)	Occurrence of the events		
	Maximum	Minimum		Date	Date	Date
Maximum value	37.8	24.6	38.2	14/10/2017	10/10/1994	13/10/1985
Minimum value	24.5	13.2	0.0	14/10/1985	12/10/1991	—

Standard Meteorological Week-42

Extreme weather Events	Temperature (°C)		Rain (mm)	Occurrence of the events		
	Maximum	Minimum		Date	Date	Date
Maximum value	38.1	32.0	154.2	16/10/2017	22/10/2010	19/10/1987
Minimum value	28.5	10.5	0.0	20/10/1987	22/10/1983	—

Standard Meteorological Week-43

Extreme weather Events	Temperature (°C)		Rain (mm)	Occurrence of the events		
	Maximum	Minimum		Date	Date	Date
Maximum value	38.3	27.2	84.0	24/10/2000	23/10/2001	28/10/1982
Minimum value	26.2	10.0	0.0	29/10/2015	28/10/1991	—

Standard Meteorological Week-44						
Extreme weather Events	Temperature (°C)		Rain (mm)	Occurrence of the events		
	Maximum	Minimum		Date	Date	Date
Maximum value	37.4	23.3	5.0	04/11/2001	30/10/1995	01/11/1982
Minimum value	26.2	10.0	0.0	29/102015	28/101981	—

Standard Meteorological Week-45						
Extreme weather Events	Temperature (°C)		Rain (mm)	Occurrence of the events		
	Maximum	Minimum		Date	Date	Date
Maximum value	35.8	22.0	16.0	06/11/2001	12/11/2010	12/11/2009
Minimum value	19.6	6.6	0.0	10/11/2008	12/11/1993	—

Standard Meteorological Week-46						
Extreme weather Events	Temperature (°C)		Rain (mm)	Occurrence of the events		
	Maximum	Minimum		Date	Date	Date
Maximum value	34.6	21.4	28.0	16/11/1986	13/11/2010	14/11/2009
Minimum value	18.4	5.1	0.0	13/11/2009	14/11/1993	—

Standard Meteorological Week-47						
Extreme weather Events	Temperature (°C)		Rain (mm)	Occurrence of the events		
	Maximum	Minimum		Date	Date	Date
Maximum value	34.0	19.4	0.0	22/11/1989	26/11/2010	—
Minimum value	23.5	1.9	0.0	24/11/2010	23/11/1993	—

Standard Meteorological Week-48

Extreme weather Events	Temperature (°C)		Rain (mm)	Occurrence of the events		
	Maximum	Minimum		Date	Date	Date
Maximum value	35.4	20.2	19.4	27/11/1986	01/12/2010	03/12/2006
Minimum value	22.5	3.5	0.0	27/11/2002	03/12/1991	—

Standard Meteorological Week-49

Extreme weather Events	Temperature (°C)		Rain (mm)	Occurrence of the events		
	Maximum	Minimum		Date	Date	Date
Maximum value	32.8	25.5	70.0	08/12/2015	08/12/2002	04/12/2013
Minimum value	20.1	1.0	0.0	10/12/1983	08/12/1990	—

Standard Meteorological Week-50

Extreme weather Events	Temperature (°C)		Rain (mm)	Occurrence of the events		
	Maximum	Minimum		Date	Date	Date
Maximum value	33.3	16.6	61.2	17/12/1985	12/12/1984	14/12/2014
Minimum value	15.6	1.8	0.0	17/12/2006	15/12/2015	—

Standard Meteorological Week-51

Extreme weather Events	Temperature (°C)		Rain (mm)	Occurrence of the events		
	Maximum	Minimum		Date	Date	Date
Maximum value	31.4	14.7	10.0	19/12/2002	22/12/2008	18/12/1986
Minimum value	18.0	2.8	0.0	21/12/1986	24/12/2015	—

Standard Meteorological Week-52						
Extreme weather Events	Temperature (°C)		Rain (mm)	Occurrence of the events		
	Maximum	Minimum		Date	Date	Date
Maximum value	31.0	19.5	16.5	24/12/1988	25/12/1991	31/12/2013
Minimum value	13.6	1.4	0.0	28/12/2003	29/12/2018	—

Source: Srivastava *et al.*, 2019,

Temperature effects interact with the soil water status which would suggest that variation in precipitation coupled with warm temperatures would increase the negative effects on crop production. These observations and the reported result suggest that more research needs to be conducted to quantify the interactions between temperature and soil water availability across germplasm within as pecies and among species to determine potential adaptation strategies to offset negative effects of extreme temperature events

Questions for chapter evaluation

Descriptive question

1. What is the difference between normal weather and extreme weather events?
2. What is the difference between global warming and climate change?
3. Write a short not on climate change.
4. What is climate variability and how climate change is different from climate variability?
5. What is greenhouse effect?
6. List the main greenhouse gases.
7. Mention the impact of climate change on crop yield.
8. What is the hot and cold wave?
9. Write the criteria for differentiating different cold and heatwaves.
10. Write a note on hot days.
11. What is drought?
12. Write different types of drought.
13. Describe the drought management in India.
14. List out the regions of India, which are drought-affected.
15. How we mitigate the impact of climate change on agriculture?

Objective questions

Fill in the blanks

i) Greenhouse gases concentration are ....with time.

ii) The major greenhouse gas emission from human activities is ....

iii) Chlorofluorocarbons (CFCs) are major contributor ....effect.

iv) Meteorological drought is a percentage deviation from ....rainfall.

v) Hot day at any place is departure of maximum temperature above ....°C.

Mention true and false against each of the questions given below.

a) The global warming associated with the increase of moisture in atmosphere.

b) The CH_4 is a major greenhouse gas.

c) The Greenhouse effect is harmful to life on the earth surface.

d) More extreme weather like droughts, heatwaves, and hurricanes are consequences associated with climate change.

e) A Cold wave is related to appreciable fall (more than 5 degree Celsius) in current temperature for their normal temperature.

f) Sea level rise is due to global warming.

g) The climate change would favour the development of fungal diseases.

h) Global warming could not affect the soil fertility.

CHAPTER 4

MANMADE DISASTER

Manmade disasters cover a wide range of events due to accidents, technical failure in any facilities, faulty design, and human negligence, misuse or escape of law resulting in huge loss of lives and property every year. Thus, it is a disaster created mainly by human negligence, faulty machine design and wrong policy design. More common manmade disasters are major fires, soil, air or water pollution, collapse of bridges and buildings, and accidents (road, rail, river, aviation marine), oil spill, bomb blast, industrial and chemical accidents etc. Man-made disasters also include nuclear, biological and chemical disasters. The impacts of these are very widespread and create human suffering, loss of life as well as long-term damage to a country's economy and productive capacity. Manmade disasters in some cases may aggravate the disastrous effects of natural hazards, such as weak house constructions in seismic zones can increase vulnerability to earthquakes.

Manmade disasters could be nuclear, biological and industrial, and some of them are very destructive and widespread, some worst manmade disasters are given below:

Nuclear : The nuclear attack on Hiroshima and Nagasaki in Japan in 1945

Biological : The use of Anthrax by terrorists to eliminate people in the USA in 2002 and COVID-19 over 195 countries of the world in 2020.

Industrial : The release of Methyl Iso-Cyanate or MIC, in Bhopal in 1984.

4.1 Nuclear Disaster

Nuclear weapons, when used caused a lot of destruction through the generation of heat, fire and radiation. The weapons such as nuclear bombs or the atom bomb are commonly called Weapons of Mass Destruction (WMD), which lead to the breakdown and collapse of social, political and economic and the food chain production systems.

Use of nuclear arms and nuclear bomb is more destructive and harmful than any other disasters in world. Several countries are using nuclear reactors for generation of power and any faulty or weak design could also create heavy loss of life and property. Two world known example are of nuclear attack on Hiroshima and Nagasaki.

4.1.1 Nuclear attack on Hiroshima

An American bomber (B-29) dropped an atombomb over the city of Hiroshima on August 6, 1945 and destruction was incredible. More than 10 square kilometers of the city was completely devastated; a City centre was vapourized (Figure 4.1) and 66,000 people were died, and 69,000 injured.

Figure 4.1 : The cloud caused by the atomic explosion over Hiroshima [*Source:* Online search free download image]

Figure 4.2 : The famous mushroom cloud over Nagasaki caused by atomic explosion [*Source:* Online search free download image]

4.1.2 Nuclear attack on Nagasaki

Another plane dropped a larger bomb, over Nagasaki on August 9, 1945 and the city was and completely devastated (Figure 4.2) and 39,000 people died, injured more than 25,000 persons.

The International Atomic Energy Agency (IAEA) is an intergovernmental agency for scientific co-operation in the nuclear field and judges nuclear accidents based on the International Nuclear Event Scale, which ranges from 1 to 7. The code 1 is considered a minor and 7, classified as a "major accident".

The Chernobyl nuclear accident is widely regarded as the worst accident in the history of nuclear power. It is the only nuclear accident that has been classified as "major accident" by the International Atomic Energy Agency.

Nuclear disasters may occur due to misuse or mishandling of Uranium mining and processing. Some events are given below:

- Accidents due to negligent handling or transportation of radioactive material.

- Accidents due to technical failure in industrial, scientific or medical facilities.
- Accidents in nuclear power stations.
- Theft of radioactive substances due to terrorism.
- Nuclear power station accidents due to natural hazards (earthquake) or aeroplane crash.
- Improper storage of nuclear waste material.

There are other causes also

- Nuclear power station disasters caused by military operations.
- Liberation of radioactive material after accidents with nuclear weapon systems.
- Ignition of nuclear strategic and tactical weapons.

Possible effects: Due to misuse and mishandling or any negligence, there may be following possible impact:

- Due to the fall-out of radioactive material the contamination/death of large parts of the population.
- Contamination of land, especially densely populated and in agricultural regions.
- Contamination of food and drinking water.
- Devastation and contamination of infrastructure.

4.2 Reactions to Nuclear Radiation

The effects of radiation on the human body can be chronic or acute. Chronic effects are those that occur some years after exposure to radiation. Examples are cancer and genetic defects. Chronic effects are of minor concern and affect any one's immediate survival in a radioactive environment.

The reaction to radiation effects are acute and of primary importance for any one survival. Some acute effects may have occurred within hours after exposure to radiation. Radiation sickness (include nausea, diarrhea, vomiting, fatigue, weakness, and loss of hair) and beta burns are examples of acute effects.

4.2.1 Protection from Nuclear Radiation and Attacks

Further radiation injuries may occur if anyone does not take proper precautions against the radiation. Don't panic in the event of a nuclear attack or radiation accident. Follow the instructions like close all doors and windows, because the radioactivity does not penetrate solid structures. Other protection are listed below:

Artificial Shelter

Use the shelters with effective shielding materials and good thickness and density. An ample thickness of shielding material will reduce the level of radiation to negligible amounts may penetrate it.

Natural Shelters

Terrain that provides natural shielding and easy shelter construction is the ideal location for an emergency shelter. Good examples are ditches, ravines, rocky outcropping, hills, and river banks. In level areas without natural protection, dig a fighting position or slit trench.

Other Shelters

While an underground shelter covered by 1 meter or more of earth provides the best protection against fallout radiation, the following unoccupied structures offer the best protection:

- Caves and tunnels covered by more than 1 meter of earth.
- Use big drains
- Basements or cellars of abandoned buildings.
- Abandoned buildings made of stone or mud.

4.3 Chemical Disaster

Chemical disasters are caused by industrial accidents, irresponsible handling of hazardous chemicals, or by their deliberate use for destruction. Large quantities of hazardous chemicals are released in soil, water and air during their installations, storages, and transportation and thus pose the greater risk for human health and their environment. Any accident involving these may have an adverse impact on both the community and the environment.

The chemical accident is an event resulting in the release of a substance hazardous to human health and/or in the environment. Such events include fires, explosions, leakages or releases of toxic or hazardous materials that can cause people illness, injury, disability or death. The extent of chemical disaster scenarios is, influenced by the military and non-military circumstance.

There are items made from in-organics materials may also release chemical in the atmosphere, water and soil. Large number (more than 400000) of chemicals are in commercial use, due to any human error in their handling or any other error will also enhance the possibility of their release in air and soil. Example: Chemical spills from oil drilling rigs. Some of the sources of chemical disasters are given below:

4.3.1 Sources of Chemical Disasters

Chemical accidents may originate in the following process:

1. Manufacturing, formulation commissioning and process operations; maintenance and disposal.
2. Material handling and storage in manufacturing units, and isolated storages (warehouses, docks and fuel depots).
3. Transportation (road, rail, air, water, and pipelines).

Chemical disasters are vulnerable due to terrorist and warfare activities also.

4.3.2 Causative Factors Leading to Chemical Disasters

Chemical disasters, in general, may result from:

i) Fire.
ii) Explosion.
iii) Toxic release.
iv) Poisoning.
v) Combinations of the above.

Major Chemical Accidents in India

1. Gas release in Bhopal Gas Disaster (1984).
2. Explosion in the Indian Petrochemicals Corporation Limited (IPCL) Gas Cracker Complex, Maharashtra (1990).
3. Explosion in the Hindustan Petroleum Corporation Limited Refinery (HPCL), Vishakhapatnam (1997).
4. A chemical released during drug manufacturing in the plant of M/s. Max India Limited, at Nanjangud in the District of Mysore (1997).
5. Leakage of ammonium carbamate solution the Urea Plant at M/s. Mangalore Chemicals and Fertilizers Limited, Panambur, Mangalore (2000).
6. Fire in an oil well in Andhra Pradesh (2003);
7. Explosion in reactor of Deccan Fine Chemicals Private Limited plant in Visakhapatnam (2014).

4.3.3 Initiators of Chemical Accidents

Several factors including human errors could spark off chemical accidents with the potential to become chemical disasters. These are:

a. Process and Safety System Failures

- Technical errors: Design defects fatigue, metal failure, corrosion etc.
- Human errors: Neglecting safety instructions, deviating from specified procedures etc.
- Lack of information: Absence of emergency warning procedures, nondisclosure of a line of treatment etc.
- Organizational errors: Poor emergency planning and coordination, poor communication with public, noncompliance with mock drills/exercises etc.
- Government should provide guidelines for the management of chemical warfare agents and chemical weapons of mass destruction to be also issued separately, which are required for ensuring a state of quick response and preparedness.

b. Natural Calamities

Natural disasters can also trigger chemical disasters. Damage to phosphoric acid sludge containment during the Odisha super cyclone in 1999 and the release of acrylonitrile at Kandla Port, during an earthquake in 2001, are some of the recent examples.

Chemical Disaster of non-military and military causes

Non-military causes

- Accidents due to negligent handling or transportation of dangerous chemical substances.
- Accidents due to technical failure in industrial, scientific or medical facilities.
- Liberation of hazardous chemical agents due to terrorism.
- Liberation of hazardous chemical agents due to natural hazards (earthquakes, floods, volcanoes).
- Accidents related to disposal measures of chemical waste.

Military cause

- Military-strikes on facilities containing dangerous chemical compounds.
- Liberation of hazardous material after accidents with chemical weapon systems.
- Employment of chemical weapons as a military action during battle.
- Accidents related to production or disposal processes.

4.3.4 Some Industry's worst disasters of the world

Oppau, Germany

In Germany on 21st September, 1921, 4,500 tonne mound of ammonium nitrate (AN) and ammonium sulfate that loosen had solidified detonated like several dynamite charges. Massive 125 meter long and 19 meter deep crater was created and the deaths of more than 500 people. The accident destroyed around 80 per cent of the homes in Oppau and ripped the roofs off houses as far as 25km away.

Texas City, Texas, US

In Texas, a fire started in French ship on 16th April, 1947 during the process of loading of ammonium nitrate fertilizer, onboard. The heat intensified and the ship exploded, killed crewmembers and onlookers. The blast was heard over 150 miles (240km) away. There were 576 people died and around 3,500 people injured.

Bhopal, India

The Bhopal Gas Tragedy is a catastrophe that has no parallel in industrial history. In Bhopal, a poisonous gas (forty tons of Methyl Iso Cyanide (MIC) leaked from Union Carbide Plant and spread throughout the sleeping city in Bhopal, India on December 3, 1984. An estimated 2500 people died, and about 300,000 suffered from unbearable injuries like breathless and blind etc.

Texas City, Texas, US

In Texas City refinery, a series of explosions occurred on March 23, 2005, when a hydrocarbon isomerization unit was restarted and a distillation tower flooded with hydrocarbons. As a result, 15 people died and more than 180 were injured.

Jilin City, China

In China a series of explosions rocked in Jilin Petrochemical's on November 13, 2005 and 05 people died and 70 were injured. Nearby equipment and storage tanks containing nitrobenzene, benzene and nitric acid feedstocks also caught fire and exploded. Benzene also leaked into the Songhua river and affected millions of people to go without drinking water.

4.4 Biological Disaster

It is a devastating effect that may be caused by an enormous spread or a sudden growth of the population of a certain kind of living organism, plants or animals (like the spread of a disease, a virus, a locust) which creates an epidemic or pandemic.

It is like an "explosion" or crowd involves a vast increase in the number of insect pests some of the most common examples are locusts, mosquitoes, butterflies, moths, beetles, bees, and dragonflies. The damage caused by locust attacks in 2019–20 in India is believed to be one of the worst (Figure 4.3). The crops were affected in more than 3.5 lakh hectares in various districts of Rajasthan and Gujarat. Crops of mustard, cumin and wheat have been devastated in the two states, affecting lakhs of farmers.

Figure 4.3 : A field crowded with locust in Gujrat in India during 2019.
[*Source:* Online search free download image]

It is a disaster that can be caused by preserving and releasing germs of deadly diseases such as smallpox, jaundice etc., or by polluting water, air with such germs in a particular area.

4.4.1 Bio-terrorism

The use of biological agents to cause death, disability or harm to human or animals population is known as bio-terrorism. It is a method of terrorist activity to prevail in mass panic and mass casualties.

Biological agents

Biological agents are living organisms or their toxic products that can kill or incapacitate people, livestock, and plants. Bacteria, viruses, and toxins are mainly used as biological agents and which could be used as weapons (Table 4.1). A small dose of 100 gram of Anthrax released over a major city may cause up to 3 million casualties.

These pathogenic organisms (smallpox and anthrax) may be used by terrorist groups as biological weapons and can cause greater harm and panic. Biological agents can be dispersed by spraying them into the air, by infecting animals that carry the disease to humans and by contaminating food and water. Though, it is difficult to grow and maintain most of the biological agents. An employee of a newspaper in Florida, died as a result of anthrax inhalation. The three basic groups of biological agents, which could be used as weapons, are bacteria, viruses, and toxins. Most biological agents are difficult to grow and maintain. Many break down quickly, when exposed to sunlight and other environmental factors, while others, such as anthrax spores, are very long lived.

Table 4.1: Common biological agents and disease

Biological agents	Disease	Infected parts
Bacillus anthracis	Anthrax	Respiratory tract
Orthopoxvirus	Smallpox, cowpox, vaccinia and monkeypox	Whole body skin
Yersinia pestis	Plague	Lymph nodes
Botulinum toxins	Botulism	Immunologically
Francisella tularensis	Tularemia	Skin or mucous membrane
	COVID-19	Respiratory tract

4.4.2 Bio-weapons and Method of delivery

The biological agents can be used as biological weapons, which are more powerful agents to mass casualties leading to civil disruptions.

- Biological agents can be disseminated with readily available technology. Simple devices such as crop dusting airoplanes or Agricultural or painting spray devices are effective delivery systems for biological agents and can be adopted to disseminate biological pathogens over great distances.
- The wind direction and speed or air movement as well as existing ventilations in building infrastructures and transportation may be used to disseminate these agents and thus to infect a large number of people.

Methods of Bio-logical agent dissemination and Delivery techniques include

- Aerosols – biological agents are dispersed into the air, forming a fine mist that may drift for miles. Inhaling the agent may cause epidemic diseases in human beings or animals.

- Animals – some diseases are spread by insects and animals, such as fleas, mice, flies, mosquitoes, and livestock.
- Food and water contamination – some pathogenic organisms and toxins may persist in food and water supplies.
- Person-to-person – spread of a few infectious agents. Humans have been the source of infection for smallpox, plague.

4.4.3 World Biggest Epidemic and Pandemic

COVID-19

Coronavirus disease (COVID-19) is an infectious disease caused by a new virus. The new coronavirus, SARS-CoV-2, which causes the disease COVID-19, has already surpassed the 2003 SARS outbreak. Now, a new respiratory illness has emerged in China. Like SARS, it's caused by a coronavirus, this one known as SARS-CoV-2. This new virus causes an illness known as COVID-19, with symptoms such as fever, cough, and shortness of breath. There's currently no specific antiviral treatment or vaccine for it. The disease causes respiratory illness (like the flu) with symptoms, such as a cough, fever, and in more severe cases, difficulty in breathing. We can protect ourself by washing our hands frequently, avoiding touching own face, and close contact (1 meter or 3 feet) with unwell people.

The 2019–20 coronavirus pandemic is an ongoing pandemic of coronavirus disease 2019 (COVID-19), caused by severe acute respiratory syndrome coronavirus 2 (SARS-CoV-2). The outbreak was first noted in Wuhan, Hubei province, China, in December 2019. The World Health Organization (WHO) declared the outbreak to be a Public Health Emergency of International Concern on 30 January 2020 and recognized it as a pandemic on 11 March 2020. As of 6 April 2020, more than 1,280,000 cases of COVID-19 have been reported in over 200 countries and territories, resulting in approximately 70,300 deaths. More than 270,000 people have recovered. The virus is mainly spread during close contact, and by small droplets produced during coughing, sneezing, or talking. These small droplets may also be produced during breathing, but rapidly fall to the ground or surfaces and are not generally spread through the air over large distances. People may also catch COVID-19 by touching a contaminated surface and then their face.

SARS (Severe Acute Respiratory Syndrome)

SARS coronavirus (SARS-CoV) – virus identified in 2003. SARS-CoV is thought to be an animal virus from an as-yet-uncertain animal reservoir, perhaps bats, that spread to other animals (civet cats) and first infected humans in the Guangdong province of southern China in 2002. An epidemic of SARS affected 26 countries and resulted in more than 8000 cases in 2003. Transmission of SARS-CoV is primarily from person to person. It appears to have occurred mainly during the second week of illness, which corresponds to the peak of virus excretion in respiratory secretions and stool, and when cases with severe disease start to deteriorate clinically. Most cases of human-to-human transmission occurred in the health care setting, in the absence of adequate infection control precautions. Within months, SARS spread to more than two dozen countries in Europe, North America, South America, and Asia. By the time the global SARS outbreak was contained, the virus spread to over 8,000 people worldwide and killed almost 800.

Ebola

The Western African Ebola virus epidemic (2013–2016) was the most widespread outbreak of Ebola virus disease (EVD) in history causing major loss of life and socioeconomic disruption in the region, mainly in Guinea, Liberia and Sierra Leone. The first cases were recorded in Guinea in December 2013; later, the disease spread to neighbouring Liberia and Sierra Leone with minor outbreaks occurring elsewhere. It is a rare, but deadly virus that causes fever, body aches, and diarrhea, and sometimes bleeding inside and outside the body. As the virus spreads through the body, it damages the immune system and organs. Ultimately, it causes levels of blood-clotting cells to drop. This leads to severe, uncontrollable bleeding. Ebola virus disease (EVD), also known as Ebola haemorrhagic fever (EHF) or simply Ebola, is a viral haemorrhagic fever of humans and other primates caused by ebolaviruses. World Health Organization (WHO) and respective governments reported a total of 28,646 suspected cases and 11,323 deaths.

H1N1

During 2009–10 the swine flu pandemic or swine flu was an influenza pandemic involving H1N1 influenza virus , albeit a new strain. First described in April 2009, the virus appeared to be a new strain of H1N1, which resulted from a previous triple reassortment of bird, swine, and human flu viruses further combined with a Eurasian pig flu virus, leading to the term "swine flu". The confirmed case around 1.4 billion and 18036 people died in this pandemic. The virus strain is Pandemic H1N1/09 virus and believed to be originated from North America.

4.4.4 Prevention and General Measures of Protection

Governments and public health authorities should identify and issue guidelines for prevention and about protection measures of biological agents.

Moreover, illnesses from biological weapons are not likely to be recognized in their initial stages and can result in widespread secondary exposure to others.

Important prevention points are given below:

1. The general population should be educated and made aware of the threats and risks associated with biological weapons.
    - Only cooked food and boiled/chlorinated/filtered water should be consumed.
    - Insects and rodents control measures must be initiated immediately.
2. Early accurate diagnosis is the key to manage casualties of biological warfare. A network of specialised laboratories should be established for laboratory diagnosis.

3. Existing disease surveillance system, as well as vector control measures, have been pursued more rigorously.
4. Mass immunization programme in the suspected area has to be vigorously followed up.
5. Enhancing the knowledge and skills of clinicians plays a vital role in controlling the adverse impact of the attack.

4.4.5 Major Biological Events across the Globe

- Use of plague bacilli by Japan in China during 1932–1945, which had caused 260,000 deaths.
- Dispersal of anthrax spores due to an accident in production unit in USSR caused 68 deaths in 1979.
- Use of *Salmonella typhimurium* by Osho followers in a restaurant (mixed in salad) in Oregaon, USA leading to 751 death in 1984.
- Use of *Shigella dysenteriae* Type 2 in Texas, USA in 1996.
- Use of Anthrax through postal envelopes in USA in October–November 2001 leading to 22 cases and 5 deaths.

4.4.6 Action Plan for Biological Disaster Management

Biological Disaster could arise from a source located either inside or outside the country. Multi-stage management plan with well integrated with the system must be required for effective management and its evaluation after post-disaster.

Pre-disaster planning in terms of system development, surveillance and post disaster should be done for following elements:

- To take notice of a situation during, which number of patients with similar health problem from a particular locality is increasing.
- Identification of focal points for control of epidemic at different levels.
- Preparation of contingency plan.
- System of information collection, active surveillance and field investigation
- Arrangement for support facilities and laboratory investigation at District, State and Central Level.

i) Emergencies preparedness and response

An integrated global alert and response system for epidemics and other public health emergencies based on strong national public health systems and capacity and an effective international system for coordinated response may be adopted. The functions of these systems are given below:

Functions

- Implementation of national capacities for epidemic preparedness and response including laboratory capacities and early warning alert and response systems;

- Support national and international training programmes for epidemic preparedness and response;
- Develop standardized approaches for readiness and response to major epidemic-prone;
- Strengthen biosafety, biosecurity and readiness for outbreaks of dangerous and emerging pathogens outbreaks (Ebola, H1N1, SARS,COVID-19);

ii) Early detection and warning signals

Early identification of an outbreak of disease is importance and their detection and advisories generation is must, as well as early warning signals should disseminated among people.

Disease outbreak response system (DORS)

When there is an outbreak resulting in the spread of an infectious disease worldwide, every country should put in place prevention and response plans.

As part of this plan, the 'Disease Outbreak Response System Condition' (DORSC) is a framework that shows the current disease situation. The framework provides us with general guidelines on what needs to be done to prevent and reduce the impact of infections.

The system takes into account:

- The current disease situation overseas
- How transmissible the disease is
- How likely it is to arrive in India
- What impact it may have on Indian community

4.5 Other Manmade Disasters

Manmade disasters are also caused by sociopolitical conflicts that rise into violence and thus much more type of disasters like riot, terrarium, war etc., are likely to happen within in a country or in many courtiers. These disasters are described below:

4.5.1 Riot

It is a violent disturbance of the peace by a crowd, public disorder caused by a group or crowd of persons, as by a crowd protesting against another group, a government policy, etc., in the streets. In a riot, group of people lashes out violence or public disturbance against authority, property or people. It is a form of civil disorder commonly characterized by a group lashing out in a violent public disturbance against authority, property or people. Riots typically involve theft, vandalism, and destruction of property, public or private. The property targeted varies depending on the riot and the inclinations of those involved. Targets can include shops, cars, restaurants, state-owned institutions, and religious buildings.

Riots often occur in reaction to a grievance or out of dissent. Historically, riots have occurred due to poor people with no jobs or living conditions, governmental oppression, taxation or conscription, conflicts between ethnic groups, (race riot) or religions (sectarian violence), or frustration with legal channels through, which to air grievances.

Types of Riots

Generally, there are four main categories of a riot:

- **Communal Riot:** In this collective violence is directed at persons of an opposing group (racial, ethnic, or religious groupings).
- **Commodity Riot:** In this collective violence is directed primarily at property, but may involve people of different groups.
- **Protest Riot:** It involves spontaneous collective violence directed against a specific policy.
- **Celebration Riot:** A celebration riot often occurs after a victory over a traditional rival or the winning of a championship.

Dealing with riots is often a difficult task for police forces. They may use tear gas to control rioters. Riot police may use less than lethal methods of control, such as shotguns that fire flexible baton rounds to injure or otherwise incapacitate rioters for easier arrest.

Followings things should be avoided during riots:

- Do not spread rumours.
- Do not violate the police instructions.
- Do not spread hatred among people.
- Do not believe and spread baseless talks appearing in social media

Following things should be done in the time of riots:

- Prevent talks, which spread rumours and violence.
- Form a peace committee in your area.
- Help the people, who are affected by the riots.
- Cooperate with police in maintaining law and order.

4.5.2 Terrorism

Terrorism is the use of violence to create a climate of fear in public to achieve its vested interest. The main characteristics of terrorism are:

- It is against the state or community.
- It has a political purpose.
- It is illegal and unlawful.
- It aims at intimidating and creating fear and panic among people.

Following are the methods of terrorism:

- Bombing
- Shooting

- Kidnappings
- Suicide attacks
- Aircraft attacks and Hijackings
- Stabbing etc.

i) Step towards prevention of terror attacks

- Inform the police about any doubtful behaviour of any persons.
- Do not touch any unclaimed thing in public places, such as shopping malls, cinemas etc.
- Cooperate in security check for safety reasons.
- If you rent your house, register it with nearest police station.
- Public places should be well equipped with CCTV cameras.

ii) Some major terrorist attract

Terrorist attack on America (9/11)

Four American planes were hijacked by terrorists' group al-Qaeda and attack different place on September 11, 2001. Two planes were crashed into the twin towers of the World Trade Centre in New York City. Third plane hit the Pentagon, the headquarters of United States department of defence. The fourth plane crashed in a field in Pennsylvania. These terrorist attacks on USA are called 9/11 terrorist attacks. 3,000 people were killed in these attacks.

Terrorist attack on Mumbai, India (26/11)

A group of Lashkar-e-Taiba terrorists from Pakistan, inturned through sea route and attack on Mumbai, India on November 23, 2008, They attacked Taj Hotel at the Gateway of India, Chhatrapati Shivaji Railway Terminus and other places in Mumbai. More than 160 people were killed and around 310 others were injured. Nine terrorists were killed and one was captured alive by Indian Security forces.

4.5.3 War

It is defined as a conflict between large groups of population, which involves physical force, violence and the use of weapons. Sometimes war is defined as an armed conflict to avoid national and international law concerning warfare.

War is of many types

i. **Civil war:** It is a conflict between two opposed parties within a country.

ii. **Cold war:** When the involvement of other parties in a civil war and receive support from external parties representing certain other political interest is called as Cold war. Like USA and Soviet Russia, Korea and Vietnam.

iii. **Guerilla war:** It is defined as a conflict of small, independent fighters or people or population who oppose a ruling government and a fight in form of small group at different time and place.

iv. **Colonial war:** It is a type of small war and relating to the various conflicts that arose as the result of overseas territories being settled by foreign powers, who had created different colony in other countries and fought during the nineteenth century between European armies in Africa and Asia.

v. **Border war:** Between two or more adjacent countries the territorial disputes often related to the possession of natural resources and also be driven by culture, religion and ethnic nationalism and caused a war between two countries is known as border war. Like India and Pakistan.

vi. **Total war:** It is a war, when unrestricted weapons were between two countries territory involved, especially one in which the laws of war are disregarded. Total war is warfare that includes all civilian-associated resources and infrastructure as legitimate military targets mobilizes all of the resources of society to fight the war, and gives priority to warfare over non-combatant needs.

vii. **World war:** A war can be termed as world war if the war spans in multiple countries on multiple continents and it involves world's most powerful and populous countries. The effect caused by war is also a determining factor to consider it as a world war. The number of death in a war is not a prime factor to term it as a world war. The world had faced two deadly world war I and II.

(a) World War I (1914–18)

The Allies (mainly Britain, France, US) won WW1, which happened from 1914–1919. Germany was the main loser, along with Austria-Hungary. The Ottoman Empire, and the other Central Powers and also Russia, although Russia withdrew from the war early due to civil war issues at home. It ended with the defeat of the Central Powers. The war was virtually unprecedented in the slaughter, carnage, and destruction it caused. The total number of military and civilian casualties in World War I, was around 40 million. There were 20 million deaths and 21 million wounded. The total number of deaths includes 9.7 million military personnel and about 10 million civilians.

(b) World war II (1939–45)

It was fought between the Axis Powers (Germany, Italy, Japan) and the Allied Powers (Britain, United States, Soviet Union, France). Most of the countries in the world were involved in some way. It was the deadliest war in all of human history with around 70 million people killed. Deaths directly caused by the war (including military and civilians killed) are estimated at 50–56 million people, while there were an additional estimated 19 to 28 million deaths from war related disease and famine. Civilian deaths totaled 50–55 million.

Questions for chapter evaluation

Descriptive questions

1. Write a note on manmade disaster.
2. Describe the type of manmade disaster.
3. What is a nuclear disaster and also describe the major nuclear disaster of world.
4. Is chemical disaster is manmade and what are causes of it?
5. What is a biological disaster? Name some of the causes of its occurence.
6. List the chemical disaster in India.
7. Mention the impact of nuclear disaster.
8. What is riot? Describe the types of riots.
9. Is terrarium is a manmade disaster? Give some examples.
10. What is bioagent and how it can be prevented?
11. Discuss the forecasting system applicable for biological disaster and its functions.
12. List the major biological disaster events of the world.
13. What is the action plan for biological disaster?
14. What is war? Write about the different types of war.
15. What is the difference between cold war and world war? Write about world war I and II.

Objective questions

a) Fill in the blanks

i. The biological agent for anthrax disaster is ....

ii. The Bhopal gas tragedy was occurred on ....

iii. Explosion in Hindustan Petroleum Corporation Limited refinery is due to..................

iv. Chemical disaster has two types .... and ....

v. Recent biological disaster, which spread over the world during 2020 is known as ....

b) Mention true and false against each of the questions given below

a. Fire is a causative factor for chemical disaster.

b. Accidents in nuclear power station may not cause nuclear disaster.

c. The second attack of nuclear bomb on Nagasaki of Japan was on 9th August 1949.

d. 26/11 is a war type disaster in India.

e. The cold war is war occurred during winter season.

CHAPTER 5

ACCIDENTS

The Manmade disaster also includes different types of accidents like road, train, air, fire etc. All these accidents are known as accidental disaster and are described in this chapter.

The Accidents are the most shocking moments for human beings and will happen anywhere and anytime. Carelessness, inexperience and attitude are associated with the occurrence of any accidents and caused heavy causality of life and properties.

An Accident is a specific, identifiable and unusual set of incidents in which the human life has a deep impact. Accidents are off various types. Most common type accidents are internal and external accidents.

Internal accident: Burns due to fires, sudden slippage, and wound etc.

External accidents: Sudden fire, road accident, air and sea accidents.

The classes of accidents are five categories:

- Structural;
- Personal injury,
- Transportation,
- Industrial;
- Scientific.

Structural accidents generally involve bridge or building collapse due to fires.

Personal injury: On account of accident there is sometimes death of the persons and in some accidents, the person suffers from serious injuries. Some of the injured persons become incapacitated for the rest of their life.

Transportation accidents most often involve roads, railways, and airlines.

Industrial and scientific involve industrial accidents like Bhopal gas disaster and explosion in chemical manufacturing plant.

5.1 Fire Accidents

Studies about occurrence of various kinds of manmade accidental disasters suggested that occurrence of a fire disaster is higher than any other accidental disaster over the world. It is also observed that fire accidents occurred very largely in number in India and over the world (Table 5.1).

Table 5.1 : Death toll due to major fire accidents in the world

S.No.	No of death	Type of building and place	Year/Date of incident
1	1995	Theater fire, Kamli, China	1893
2	1670	Theater fire ,Canton, China	May 25, 1845
3	900	Theater fire Shanghai, China	June 1871
4	800	Theater fire, St. Petersburg, Russia	1836
5	667	Cinema fire ,Xinjiang, China	1977
6	538	School fire, Mandi Dabwali, India	December 23, 1995
7	448	Theater Fire, Vienna, Austria-Hungary	December 8, 1881
8	430	Circus Ferroni fire Berdichev, Russia	January 13, 1883
9	422	Cinema Rex fire, Abadan, Iran	August 20, 1978
10	396	Supermarket fire, Asunción, Paraguay	August 1, 2004
11	360	prison fire, Comayagua, Honduras	February 14, 2012
12	59	Uphaar Cinema fire, Delhi, India	June 13, 1997
13	24	Nakumatt supermarket fire, Nairobi, Kenya	January 2009
14	22	Surat fire	2016
15	9	Nagaland Forest fire ,India	24 May, 2019

Fire hazard: It is a situation which increases the likelihood of a fire may start or may slow down the escape mechanism in the event a fire occurs. In other words threats to fire safety are referred to as fire hazards. Moreover, the amount of damage from fire hazards over a period of time have often been more than the damage caused by natural hazards such as cyclones or floods.

5.1.1 Building Fire

It is a type of high speed combustion process in which all building materials are rapidly oxidized and released large quantity of smoke, light and heat, and damage the building property and life.

5.1.2 Coal Fire

Coal fire is simply burning of huge amount of coal in a coalfield or storage and it refers to burning of coal stratum, coal storage pile or coal waste pile (Figure 5.1). Spontaneous combustion is the most common cause of coal fires. In this exothermic reaction process oxygen accumulates on surface of coal and due to high temperature (above 80°C) the coal can ignite and starts burning.

Figure 5.1 : Coal fire in Jharkhand, India. [*Source:* Online search free download image]

Global Coal Fire Occurrence

Bulk storage of any combustible materials lead to fire risk in many large storage areas such as waste bunkers, wood or paper stockpiles and coal storage yards. The fires usually start as 'hot spots' in the coal accumulation. These are places, where the generated heat cannot be dissipated efficiently, while there is still enough oxygen to promote the oxidation reaction of the coal. An uncontrolled coal fire is an environmental and economic problem and occurs in many countries worldwide including India, China, Russia, the United States, Indonesia, Venezuela, Australia, South Africa, Germany, Romania and the Czech Republic etc.

5.1.3 Forest Fire/Wildfire

Forests are extremely important to the survival of human beings on earth. Forest fire an uncontrolled fire that occurred in the wilderness or in forest area (Figure 5.2). It can be caused by many factors like lightning, volcanic eruptions and also human actions. At the global level, forest fires are common in Australia, Western Cape of South Africa, the hills of South Asia and the forested areas of the United States and Canada. In South East Asia, forest fires are considered one of the biggest threats to the forest resources. Some of the south east Asian regions are prone to forest fire and a few area are in India, 50 percent forest area is prone to fire in Bhutan, approximately 40 per cent of the total forest area, are most susceptible to frequent forest fires. Nepal is yet another South Asian country, highly vulnerable to the forest fires.

5.1.4 Wildfire

A wildfire is any uncontrolled fire in combustible vegetation in a wilderness area. Wildfires are characterized in terms of the cause of ignition, speed of propagation, the combustible material, and the effect of weather on the fire. It occurs when combinations of conditions are right, drought and high wind combined. Other names, such as brush fire, forest fire, grass fire, hill fire, peat fire, vegetation fire and wildland fire may be used to describe the same phenomenon depending on the type of vegetation being burned. Wildfires occur on every continent except Antarctica. Wildfires can cause extensive damage, both to property and human life, but they also have various beneficial effects on wilderness areas. Some of the countries are prone to forest fires like Australia and the United States, but fires also occur fairly regularly in Europe, Indonesia, and Greece.

Figure 5.2 : Forest fire in Uttarakhand in India
(*Source:* Online search free download image)

Strategies of wildfire prevention, detection, and suppression have varied over the years, and wildfire management experts encourage further development of technology and research. One of the more controversial techniques is *controlled burning*: permitting or even igniting smaller fires to minimize the amount of flammable material available for a potential wildfire. While some wildfires burn in remote forested regions, they can cause extensive destruction of homes and other property located in the wild and-urban interface: a zone of transition between developed areas and undeveloped wilderness.

5.1.5 Causes of Forest/Wildfires

Lightning, human carelessness, slash-and burn farming, volcano eruption, underground coal fire. Heatwaves, droughts, and excessive dryness, hot summer. Forest or bush fire is a major hazard for forest cover in the country. As per Forest Survey of India report, 50 per cent of the forest cover of the country is fireprone, out of which 6.17 per cent is prone to severe fire damage causing extensive loss to forest vegetation and environment. Average annual physical loss due to forest fire in the country is estimated to worth Rs. 440 crores (FSI,2003). The coniferous forests in the Himalayan region are very susceptible to fire and every year there are one or more major fire incidences in these areas. The other parts of the country dominated by deciduous forest are also damaged by fire up to an extent. It is worth mentioning that in India 90 per cent of the forest fires are man-made.

5.1.6 Major Forest Fires Events

The forest fire records since 1900 AD showed many catastrophic forest fires affecting a large number of human population all over the world. Historically, there have been some major forest fires events, which made considerable impacts on lives and properties. The world's greatest forest fires disaster in terms of people died and affected are presented in table 5.2 below:

Table 5.2 : World most catastrophic forest fires (1900–2007)

Country	Date	Death/Total Affected
U.S.A., Boston, Massachusetts (Cocoanut Grove fire)	November 28, 1942	492 killed
Indonesia	November 1997,	200 killed
Greece	2007	50 killed
Indonesia, Forest fire	October, 1994	30,000,00 affected
Macedonia FRY, Forest fire	July, 2007	10,00,000 affected
United States, Scrub/grassland fire	21–10–2007	640,064 affected
Argentina, Forest fire	22–01–1987	152,752 affected
Portugal, Forest fire	Jan., 2003	150,000 affected
Paraguay, Forest fire	September, 2007	125,000 affected
Russia, Forest fire	20–07–1998	100,683 affected
China P Rep, Forest fire	May, 1987	56,313 affected
Nepal, Forest fire	Mar, 1992	50,000 affected
Myanmar, Forest fire	11–04–1981	48,588 affected

[*Source:* EMDAT]

i) Forest fire proneness in India

According to an assessment of the Forest Protection Division of the Ministry of Environment and Forests, Government of India, 3.73 million hectares of forests are affected by fires annually in India, which is 50 percent of the total forest area. The forest fire incidence is normally high during the dry winter months. Freezing temperatures and lack of rainfall are responsible for drying of perennial grasses, and increasing wind velocity quickens the drying process, thereby making the grass covered area flammable. The forest fire prone states are Himachal Pradesh, Assam, Arunachal Pradesh, Nagaland, Mizoram, Chhatishgarh and Madhya Pradesh.

ii) Forest fire in Himachal

Forest fires are recurring phenomenon in Himachal Pradesh. The forests fire were occurred in the summers and the fire season starts from mid-April. A forest fire occurred in June 2007 and destroyed hundreds of hectares of forest in Himachal Pradesh. After the heavy loss of forest due to this forest fire, the state forest department of Himachal Pradesh has decided to implement the satellite based Indian Forest Fire Response and Assessment System and established this system in 2007.

Recently there was another forest fire occurred in February 2014 in Kullu; which destroyed pine and deodar forest. Long dry spell was the major reason for this forest fire.

iii) Forest fire in Nagaland

A forest fire occurred in Mount Japfu in Nagaland in February 2014 and destroyed many hectares of forest cover. The Indian Air force has engaged its Mi-17 V5 helicopters to fight this major forest fire.

5.1.7 Indian Forest Fire Response and Assessment System (INFFRAS)

Indian Forest Fire Response and Assessment System (INFFRAS) was established by the Indian Government at the National Remote Sensing Centre (NRSC), to facilitate forest fire monitoring and management. The system, developed by the National Remote Sensing Agency (NARSA) will provide fire alerts to the department in 'near real-time' based on satellite data received from the moderate resolution imaging spectra radiometer (MODIS). It captures both visible and infrared wavelengths to helping detecting forest fires. This system accurately locate the site of a fire, the system provides satellite imagery, besides latitude and longitude.

The INFFRAS has been set up to preliminarily carry out:

- Active Forest Fire Monitoring.
- Burnt Area Assessment.
- Fire burnt area progression monitoring.
- Ecological damage assessment due to forest fires.

INFFRAS is providing fire-related inputs to the user departments in terms of fire information along with value-added services such as forest administrative boundaries, vegetation type burnt, nearby villages, etc.

5.1.8 Oil Fire

Petroleum and its byproducts are highly inflammable. Due to human error/accidents or natural events (lightning) a fire is ignited in oil in presence of high pressure and temperature called oil fire (Figure 5.3). It can occur at a small scale in oil tanks/field or on a huge scale in oil wells.

Figure 5.3 : Oil fire in oil well.

i) Effects

Oil well fires/tanks can cause the loss of millions of barrels of crude oil per day and create an ecological problem due to large amounts of smoke and unburnt petroleum falling on earth. Smoke from burnt crude oil contains many chemicals, including sulfur dioxide, carbon monoxide, soot, benzopyrene, Polyaromatic hydrocarbons, and dioxins, which are harmful to human health and its environment.

Example- Oil tanks fires was seen in Jaipur and oil well fire generally seen in Kuwait, and can cause enormous economic losses.

ii) Protection and prevention:

The techniques include:

- Dousing with copious amounts of water.
- Liquid nitrogen or water is then forced in at the bottom to reduce the oxygen supply and put out the fire.
- Drill relief wells into the producing zone to redirect some of the oil and make the fire smaller.
- Using a gas turbine to blast a fine mist at the fire. Water is injected to the compressor section of the turbine in large quantities.
- Using dynamite to 'blow out' the fire by blasting fuel and oxygen from the flame and consuming oxygen in the combustion.
- Dry Chemical (mainly Purple K) can be used on fires, such as those in refineries.

iii) Major oil fire in India

Kanpur oil fire on 26th April, 2014.

5.1.9 Fire Safety

Fire safety is an alert or precautions those are taken to reduce the likelihood of fire, which results in lowering the loss of life and property. Every planned construction of a building should have taken the precautionary measure for safe removal of inhabitant and their belonging in case of fire. The owner or construction company should have to take fire safety certificate before the transfer of occupation of the building. Every habitat in an area should be obtained the simple Do's and Don't certificate to create awareness on to reduce the fire accidents in case of fire respond effectively.

i) Fire safety is often a component of building safety and includes

- Not exceeding the maximum occupancy within any part of the building.
- Maintaining proper fire exits and proper exit passage.
- Placing and maintaining fire extinguishers in easily accessible places.
- Proper storage for hazardous materials inside the building
- Prohibiting flammable materials in certain areas of the facility.

- Periodically inspecting buildings for violations and issue order for its compliance.
- Maintaining fire alarm systems for detection and warning of fire.
- Maintaining a high level of training and awareness of occupants and users of the building to avoid obvious mistakes, such as the propping open of fire doors.
- Conduct public fire safety education and fire drills at regular intervals throughout the year.

ii) Key elements of a fire safety policy

Built the building following the local building code and maintain the facility in the building for fire safety and people of the building should aware about the facilities available in building and also aware of the fire safety. All the passage and staircase in a building map must be displayed on every floor that map may be followed for evacuation from building in case of fire accidents.

iii) Some common fire hazards

- Blocked cooling vent.
- Overloaded electrical system.
- Fuel store areas with high oxygen concentration or insufficient protection.
- Materials that produce toxic fumes when heated.
- Objects that block fire exits.
- Combustibles near or around the clothes dryer.
- Incorrectly installed wiring.
- Misuse of electrical appliances.
- Lit candles left unattended.
- Improperly-extinguished tobacco.
- Failure to clean and maintain the clothes dryers exhaust duct.
- Combustible solutions on clothes placed in the clothes dryer.
- Flammables left near a hot water heater.
- Fireplace chimneys not properly or regularly cleaned.
- Misuse of wood burning stoves.

Improper use and maintenance of gas stoves often create fire hazards

iv) Fire code

The fire code is a model code adopted by the state or local jurisdiction and enforced by fire prevention officers within municipal fire departments. It is a set of rules prescribing minimum requirements to prevent fire and explosion hazards arising from storage, handling, or use of dangerous materials, or from other specific hazardous conditions. It complements the building code.

The fire code is aimed primarily at preventing fires, ensuring that necessary training and equipment will be on hand, and that the original design basis of the building, including the basic plan set out by the architect, is not compromised. The fire code also addresses inspection and maintenance requirements of various fire protection equipments in order to maintain optimal active fire protection and passive fire protection measures. A typical fire safety code includes administrative sections about the rule-making and enforcement process, and substantive sections dealing with fire suppression equipment, particular hazards such as containers and transportation for combustible materials, and specific rules for hazardous occupancies, industrial processes, and exhibitions.

v) Public fire safety education

Fire prevention programs may include distribution of smoke detectors, visiting schools to review key topics with the students and implementing nationally recognized programs. Other programs or props can be purchased by fire departments or community organizations. These are usually entertaining and designed to capture children's attention and relay important messages and demonstrate the mix audio messages and visual queues with hands-on interaction.

Fire prevention programs like "Risk Watch" and "Learn not to burn" may be demonstrated in public places and schools to create awareness among people.

Other programs or props including audio and visual messages with hands-on interaction may be purchased and demonstrated in public by fire departments or community organizations. These programs will be helpful to create awareness about an escape in case of fire.

vi) Protection and prevention

Fire prevention is intended to reduce sources of ignition. Fire prevention also includes education to teach people how to avoid causing fires. Buildings, especially schools and tall buildings, often conduct fire drills to inform and prepare citizens on how to react to a building fire. Fire stations are to be established in well organized areas to put out uncontrolled fires. Keep and maintain the fire equipments for use of trained firefighters, fire bus with availability of fire preventing materials like water; foam etc. must be provided at all fire stations in case of fire.

Fire fighting services are provided in most developed areas to extinguish or contain uncontrolled fires. Trained firefighters use fire apparatus, water supply resources such as water mains and fire hydrants foam depending on what is feeding the fire. Fire prevention is intended to reduce sources of ignition. Fire prevention also includes education to teach people how to avoid causing fires.

Building Fire Protection (BFP): Modern buildings built under the strict design and buildings codes of today have many fire protection systems installed by default. These systems assist with detection and response to fire related emergencies.

vii) House fire

House fires killed and injured many people every year and damaged the valuable properties. Here are some steps that may be learned, which may lessen the chance of damage due to fire. The following steps to remember:

- Inspect your home at some interval of time.
- Stay in the kitchen when using the range for cooking
- Don't cook when drinking alcohol, using drugs, or are very tired.
- Don't sit down or lie down when smoking.
- Check the condition of your electrical fitting in the house.

5.2 Deforestation

Deforestation is the process of cutting the forests and convert it into arable land, wasteland, industrial or urban area. The loss of forests also causes desertification. The roots of trees dig deep into the ground, penetrating several layers. They hold together these layers and prevent the formation of dust and thus maintain the topsoil intact. In the absence of trees, dust is formed and heavy rainfall and high sunlight damage the topsoil in clearings of the tropical rainforests. Heavy rainfall with heavy winds and storms has flown away from the top soil and thus the availability of fertile land decreases. Therefore deforested areas appear like desert.

5.2.1 Forests and Disaster Vulnerability

Forest ecosystems play an important role in reducing the vulnerability of communities to disasters. It is not only reducing their physical exposure to natural hazards, but also providing them with the livelihood resources. Vulnerability is intensifying around the world due to the degradation of forest ecosystems. Numerous examples are available across the world – The super cyclone hit Odisha coast in October 1999 and killed 9000 people, is one of the examples of deforestation and disaster vulnerability. The Indian Ocean tsunami, 2004 had seen the deforestation-disaster vulnerability link. Government may frame a strict policy for forest conservation and promote the tree planting across the communities.

5.3 Road Accidents

A road accident refers to any accident involving at least one road vehicle, occurring on a road open to public circulation, and in which at least one person is injured or killed. India has a vast road network of 5.9 million km of which the expressways account for 1.3 lakh km. The number of vehicles has been growing at a rapid pace of 12 per cent per annum since the eighties and consequently, traffic on the roads is growing at 7 to 10 per cent per annum. The road accidents have occurred mostly due to overtaking, overloading and high speed vehicles and crossing on road. Such accidents could occur at any time and any place as well as and often involve multiple injuries or deaths.

The major causes of accidents are due to rapid increase of vehicles, bad roads, untrained drivers, slackness of traffic police, lack of obeying the traffic rules etc. The citizens of India are more careless about their own safety and quite frequently meet with accidents. In India, 1.5 lakh people are died and around 5.0 lakh people injured by the road accident in every year. Road accidents in India are highest in the world.

> *As per 2017 report on road accidents in India, it was observed that 1214 road accidents occurred in every day and 17 person died every hour.*

The road accidents occurred mostly on national highways and expressway in India. The road accident prone zone of India is described below:

Here are some facts about India's most accident-prone roads.

The top five cities of most road accident prone of India are Chennai, Delhi, Bengaluru, Indore and Kolkata.

Top five states Uttar Pradesh recorded the maximum number of road deaths (17,666), followed by Tamil Nadu (15,642), Maharashtra (13,212), Karnataka (10,856) and Rajasthan (10,510).

The most road accident prone highways of India are given in Table 5.3.

Table 5.3 : Most accident prone highway of India

S. No.	Name of Highway	Length in Km	Main connecting cities/states	Number of black spots
1	Delhi – Kolkata Highway	Spreads over 1,465 km	It connects Delhi, Haryana, Uttar Pradesh, Bihar, Jharkhand, and West Bengal	59 black spots
2	Delhi – Mumbai Highway	Spreads over 1250 km	It connects the cities – Delhi, Faridabad, Jaipur, Vadodara, and Mumbai	45 black spots
3	Nongstoin – Sabroom Highway	Spreads over 723 kilometres	It connects the states Meghalaya, Assam, and Tripura.	38 balck spots
4	Thane – Chennai Highway	Spreads over 1,235 kilometres	It connects Mumbai, Pune, Bangalore, and Chennai	27 black spots
5	Chennai – Theni Highway	Spreads over499 km	It connects Chennai to Theni	24 black spots

To reduce the road accidents on the above highways, government may take the necessary steps at least reduce the black spots on these highways. The construction of over foot-bridges and subways will also save precious human lives.

Road accidents result in a major loss to the economy. The estimated GDP loss due to road traffic crashes amounted to three percent.

5.3.1 Causes of Road Accidents

The road accidents occur due to fault of drivers, lack of pedestrian, mechanical defect in vehicles, bad conditions of road, bad weather and other reasons like non-functioning of signals, absence of reflectors, cattle/animals on the road during night etc. The lack of knowledge of traffic rules to drivers and road users, over speeding, overloading, drunken driving are the major reasons of road accidents in the country.

Safety rules to avoid accidents:

1. Drivers should keep to the vehicle in left side, allow the traffic in the opposite side to pass.
2. Try to avoid the overtake.
3. Do not raise the speed when being overtaken by another vehicle.
2. Slow down the vehicle when passing the road junctions, railway crossing, and villages/town.
3. Drive slowly when passing a procession, repairs of the roads, near the schools.
4. Stop the vehicle when passing the zebra crossing by the pedestrians.
5. Drive the vehicle with the speed limit as per issued by the Motor vehicle act.
6. Always obey the traffic or safety rules.

Government steps for minimizing road accidents:

1. Provide assistance in setting up the Driving schools.
2. Provision of refresher courses to drivers or general public of heavy motor vehicles.
3. More stringent in issuing licenses.
4. Reduce the number of vehicles on the roads.
5. Be strict about usage of helmets.
6. Make separate lanes for heavy vehicles.
7. Awareness campaign on road safety rules among road users through audio-visual, print media.
8. Speed controlling measures, such as speed bumps, rumble strips, road markings, traffic signs, and roundabouts.
9. School children need to be imparted road safety education, specifying the various safety measures to be adopted while on the road.
10. Government should care for the improvements in road and facilities for road users on highways.
11. Below 18 years of age, no one should allow to drive any vehicle on road and strict rules should be followed by every citizen.

5.4 Rail Accidents

Indian Railways, over 67,368 km long, is the world's fourth largest network (123,236 km) after the US, Russia and China. Considering the huge number of passengers, the frequency of travel and the long distances covered, rail transport is considered as safe transportation mode in India. The rail accidents may be caused by human or system failure, which may affect normal movement of rail services and cases of loss of human life and property. An analysis of the accident statistics reveals that derailments constitute a majority of the accidents followed by unmanned level crossing accidents. The network of Indian railway is maintained and operated by Indian government. The Indian rail network comprises a route of 67,368 km and 7,349 stations. It is the fourth-largest railway network in the world and forty nine percent of the routes are electrified, while thirty-three percent of them are double or multi-tracked. It is one of the busiest networks in the world, transporting 8.107 billion passengers and over 1.108 billion tonnes of freight annually. The Indian Railways is among the world's largest rail networks.

In Railways, disaster is defined as a major train accident leading to heavy causalities and disruption to traffic for a long period. Train accidents are further classified as consequential train accidents and other train accidents. Consequential train accidents include train accidents having serious repercussion in terms of loss of human life/human injury/loss of Railway property/interruption to rail traffic. Some worst accidents occurred in India where trains have collided, gone off the tracks or fallen in the river.

5.4.1 Major Train Accidents (since 2000)

A large number of railroad accidents take place in India. Every other day, people get injured or die in small accidents. Larger railway accidents including collisions and derailments cause huge losses to life and inflict injuries and disabilities to many more. The major rail accidents since 2000 are given Table 5.4; and biggest train accidents of India in Table 5.5.

Table 5.4 : Major train accident in India

S. No.	Date/Year	Train name	Place	Loss of life	Injured
1.	02 Dec., 2000	Howrah-Amritsar Mail	Banjara and Sadhugarh in Punjab	46 people	150 injured
2.	22 June, 2001	Mangalore-Chennai Mail	Kozhikode in Kerala	42 people	300 injured
3.	27 February, 2002	Sabarmati Express	Godhra station in Gujarat	48 people	43 injured
4.	12 May, 2002	New Delhi-Patna Shramjeevi Express	Jaunpur in Uttar Pradesh	12 people	25 injured

S. No.	Date/Year	Train name	Place	Loss of life	Injured
5.	04 June, 2002	Mathura-Kasganj Express	Medu railway station in Hathras	24 people	07injured
6.	10 September, 2002	Kolkata-New Delhi Rajdhani Express	Biha	120 people	37 injured
7.	22 June, 2003	Karwar-Mumbai Central	Vaibhavwadi station in Maharashtra.	43 people	25 were injured
8.	16 June, 2004	Matsyagandha Express	Maharashtra's Raigarh distric	20 people	>60 injured
9.	1 December, 2006	—	Bihar's Bhagalpur district	35 people	injuring 17
10.	19 July, 2010	Bhagalpur-Ranchi Vananchal Express	Birbhum district of West Bengal	89 people	51 injured
11.	14 November, 2009	Mandore Express	Bassi town near Jaipur	07 people	>60 injured
12.	21 October, 2009	Goa Express	Mathura-Vrindavan in Uttar Pradesh.	22 people	26 injured
13.	28 May, 2010	Gyaneshwari Express	West Midnapore district in West Bengal	148 people	87 injured
14.	22 May, 2012	Banglore-bound Hampi Express	Anantapur district in Andhra	25 people	32 injured
15.	08 May, 2014	A collision between train and a Bulero car	Maharajganj in Uttar Pradesh	13 people	nearly 46 were injured
16.	4 May, 2014	Sawantvadi Passenger	Diva Junction	20 people	100 injured

S. No.	Date/Year	Train name	Place	Loss of life	Injured
17.	20 March, 2015	Dehradun-Varanasi Janta Express	Rae Bareli in Uttar Pradesh	48 people	150 injured
18.	20 November, 2016	Pukhrayan train derailment	Pukhrayan, in Uttar Pradesh.	170 people	>260 injured
19.	21 January, 2017	Kuneru train	near Kuneru, V izianagaram	41 people	68 injured
20.	19 October, 2018	A train ran into a crowd of spectators	Amritsar, Punjab	49 people	100 injured

Table 5.5 : Biggest disaster by train accidents

S. No.	Date/year	Train/place	Loss of life	Injured
1.	17 July, 1937	An express train/near Bihta station, Partna, Bihar	119 people	180 injured
2.	28 September, 1954	A train crashed into the Yasanti river south of Hyderabad	139 people	>100 injured
3.	2 September, 1956	A bridge collapsed under a train Mahbubnager near Hyderabad	125 people	222 injured
4.	14 July, 1969	A freight train crashed into a stationary passenger train at Jaipur	85 people	130 injured
5.	6 June, 1981	a train fell into the river Bagmati in Bihar	between 300 and 800 people	
6.	15 September, 1984	Jabalpur Gondia passenger Balaghat, Madhya Pradesh	150 people	85 injured
7.	8 July, 1988	Island Express fell into Ashtamundi lake at Bangalore–Thiruvananthapuram	105 people	200 injured
8.	20 August, 1995	Purushottam Express near Firozabad	400 people	250 injured
9.	26 November, 1998	Jammu Tawi–Sealdah Express at Khanna	212 people	150 injured

S. No.	Date/year	Train/place	Loss of life	Injured
10.	2 August, 1999	Awadh-Assam Express at Gaisal in North Frontier Railway's Katihar division	268 people	359 injured
11.	9 September, 2002	Howrah-New Delhi Rajdhani Express Dehri-on-Sone stations, Bihar	140 people	55 injured
12.	29 October 2005	Repalle Secunderabad delta passenger Nalgonda district of Andhra pradesh	100 people	25 injured
13.	26 November, 2005	Mandovi express were crushed in a tunnel near Ratnagiri after a rockfall.	100 people	500 injured
14.	10 July, 2011	Kalka Mail derailed near Fatehpur, Uttar Pradesh	70 people	>300 injured
15.	20 November, 2016	Indore–Rajendra Nagar Express Pukhrayan in Uttar Pradesh	152 people	> 260 injured

5.5 Air Accidents

The data reported from the air accidents all over the world indicates that the deaths due to air accidents are fewer as compared to any other modes of transport and hence, the air travel is considered the safest mode of transportation. There have been a few major air accidents over the Indian skies.

5.5.1 Causes of Plane Crash/Air accident

Though air travel is one of the safest forms of transportation, accidents do happen with dramatic and terrifying results. The causes of these aviation accidents vary greatly depending on specific circumstances and problems that may develop during the flight process. However, the major causes are listed below:

- Piloting errors
- Faulty equipment
- Violations of FAA regulations
- Design or structural problems
- Flight service negligence
- Air traffic controller error
- Third party carrier selection negligence
- Maintenance or repair negligence
- Fueling error.

Other causes of aviation accidents include bird hazards, mid-air collisions, air traffic control errors, structural defects, lack of maintenance, air show accidents, and search and rescue operations. These incidents can be completely avoided through careful preparation and effective safety techniques.

5.5.2 Air Accidents due to Inclement Weather

Sometimes Aircraft accidents occur due to bad weather. Although poor weather conditions are beyond the control of pilots, airlines, and flight crew, they have a responsibility for the safety of their passengers and follow the weather advisories.

Pilots should not fly in hazardous weather conditions such as blizzards, turbulence, poor visibility, high temperature and wind etc. Rain and thunderstorms can be extremely hazardous to aviation. Turbulence, cumulus clouds, high winds, ice, hail, lightning, loss of visibility, electrostatic discharge, tornadoes, altimetry errors, and wet runways often accompany rain and must be managed by pilots and flight crews. In most situations, pilots are instructed to avoid severe thunderstorms and rain due to the risks they may pose for passengers and crew.

5.5.3 List of Air Accidents

The major air accidents were happened in India as well as in the world. The chronology of air accidents in India and world are given in table 5.6 and 5.7 respectively.

Table 5.6: Major air accident in India

S.No.	Date	Aircraft	Place	Loss of life
1.	Jul 7, 1962	Alitalia flight from Sydney	Mumbai	94 people
2.	Jan 1, 1978	Air India flight	Arabian Sea	213 people
3.	Jun 21, 1982	Air India flight	Mumbai	17 people
4.	Oct 19, 1988	Air India flight	Ahmedabad	124 people
5.	Feb 14, 1990	Air India flight	Bangalore	92 people
6.	Aug 16, 1991	Air India flight	Imphal	69 people
7.	Apr 26, 1993	Air India flight	Aurangabad	55 people
8.	Nov 12, 1996	Saudi Arabian Airlines	Charki Dadri in Haryana	349 people
9.	July 17, 2000	Alliance Air flight	Patna	56 people
10.	May 22, 2010	Air India	Mangalore	166 people

Table 5.7: Major air accident of the world

S.No.	Date	Aircraft	Place	Loss of life
1.	May 6, 1937	The Zeppelin Hindenburg	New Jersey, United States	35 people
2.	November 8, 1940	A Deutsche Luft Hansa Junkers Ju	Saxony, Germany	29 people
3.	July 28, 1943	American Airlines Flight	Kentucky, United States	22 people
4.	February, 10 1944	American Airlines Flight	Mississippi River in the United States	24 people
5.	November 14, 1946	KLM Douglas DC	Amsterdam in Netherlands	26 people
6.	May 30, 1947	Eastern Air Lines Flight	Maryland, in the United States	53 people
7.	August 31, 1950	TWA Flight	Cairo, Egypt	55 people
8.	November 3, 1950	Air India Flight	Mont Blanc in France	48 people
9.	February 3, 1959	American Airlines Flight	New York, United States	68 people
10.	September 12, 1961	Air France Flight	Rabat-Salé Airport	77 people
11.	June 3, 1962	Air France Flight	Atlanta, Georgia	130 people
12.	February 8, 1965	Eastern Air Lines Flight	New York, United States	84 people
13.	May 20,1965	PIA Flight	Cairo International Airport	119 people
14.	February 4, 1966	All Nippon Airways Flight	Tokyo Bay, Japan	133 people
15.	July 30, 1971	All Nippon Airways Flight	Morioka, Japan	162 people
16.	January 22, 1973	a Boeing 707	Kano International Airport in Nigeria	176 people
17.	May 31, 1973	Indian Airlines Flight	Palam Airport in New Delhi	51 people
18.	September 19, 1976	Turkish Airlines Flight	Karatepe, Turkey	157 people
19.	October 12, 1976	Indian Airlines Flight	Bombay Airport	98 people

S.No.	Date	Aircraft	Place	Loss of life
20.	April 25, 1980	Dan-Air Flight	Tenerife, Spain	142 people
21.	June 8, 1982	VASP Flight	hillside in Brazil	137 people
22.	June 23, 1985	Air India Flight	International Airport Heathrow, London	329 people
23.	August 12,1985	Japan Airlines Flight	Mount Osutaka	520 people
24.	January 8, 1996	Air Africa crash	Democratic Republic of the Congo	297 people
25.	February 6 ,1996	Birgenair Flight	crashes into the ocean off Puerto Plata in the Dominican Republic s	189 people
26.	October 31, 1999	Egypt Air Flight	Atlantic Ocean off Nantucket, Massachusetts	217 people
27.	March 8,2014	Malaysia Airlines Flight	Plunged into the Indian Ocean, killing all on board.	239 people

Ten Major Air Accidents of India

- *Turkish Airlines Flight 981, March 3, 1974; Fatalities: 346.*
- *The Tenerife Airport Disaster, March 27, 1977. Fatalities: 583.*
- *American Airlines Flight 191, May 25 1979. Fatalities: 273*
- *Saudia Flight 163, When : August 19, 1980; Fatalities: 301.*
- *Air India Flight 182, June 23, 1985. Fatalities: 329.*
- *Japan Airlines Flight 123, August 12, 1985. Fatalities: 520*
- *Iran Air Flight 655. July 3, 1988. Fatalities: 290.*
- *Charkhi Dadri mid-air collision, November 12, 1996. Fatalities: 349.*
- *Malaysia Airlines Flight 370, March 8, 2014. Missing: 275.*
- *Iranian Air Force Ilyushun II-76, February 19, 2003. Fatalities: 275.*

5.6 Industrial accidents

Generally, involve the chemical or petroleum industries. Industrial accidents seem to occur in bunches. For example, in 1984, three (3) major industrial disasters took place in India, Brazil, and Mexico, all at about the same time. over the spillover effects from industrial disasters began shortly after 1976. Some of the accidental disasters are given below:

Bhopal Accident 1984, India

December 3, 1984, methyl isocyanate gas leaked from the Union Carbide facility during the early hours of the morning while local residents slept. Around 2,000 people died immediately, with another 8,000 dying later.

Flixborough, UK – June 1, 1974

In 1974, cyclohexane vapour leaked from ruptured pipework at the Nypro (UK) site at Flixborough. This resulted in an explosion that killed 28 people and injured 36. Offsite, 53 injuries were reported. Property in the surrounding area was also severely damaged.

Seveso, Italy – July 10, 1976

On July 10, 1976, in a small Italian town Seveso, a reactor at the ICMESA chemical plant overheated, resulting in an explosion and highest known exposure of dioxins in a residential area. A toxic cloud containing 2,4,5–Trichlorophenol, used to make pesticides and antiseptics, spread to the densely populated city of Seveso.

Toulouse, France – September 21, 2001

A explosion at Atofina's Grande Paroisse fertilizer plant in Toulouse, France was observed in the year 2001. Around 300 tonnes of ammonium nitrate exploded, destroying the site and wrecking buildings 3km away in the city centre. It was responsible for the death of 30 people, and 10,000 injuries.

Schweizerhalle, Switzerland – November 1, 1986

A major fire at the Sandoz chemical factory in 1986 washed chemicals into the river Rhine, one of Europe's busiest waterways. The spill caused severe pollution, which took years to eradicate and killed an estimated 500,000 fish.

Besides the above accidents, other industrial accidents are shown in table 5.8.

Table 5.8: Industrial accidents in India

S. No.	Date/year	Name of accidents	Death toll
1.	10 March 1906	Courrières mine disaster, Courrières, France	1,099
2.	15 December 1914	Mitsubishi Hōjō, Kyūshū, Japan	687
3.	1942	Benxihu Colliery explosion, China,	1,549
4.	April 16 & 17, 1947	Texas City Disaster (Texas City, Texas	568

S. No.	Date/year	Name of accidents	Death toll
5.	9 May, 1960	Laobaidong colliery coal dust explosion Datong, China	682
6.	14 September, 1997	Hindustan Petroleum Refinery fire, Vishakapatnam, Andhra pradesh, India	22
7.	21 November, 2006	Methane explosion in Halemba coal mine, Ruda Śląska, Poland	23
8.	December 24, 2008	Ganglu Iron and Steel Co Ltd., gas leak (Hebei, China,)	17
9.	August 10, 2009	Handlová mine blast, Handlová, Slovakia	20
10.	April 20, 2010	Robin Hood Flour Company grain storage elevator explosion, Iowa, USA	11
11.	April 2, 2010	Tesoro Refinery, Anacortes, Washington	7

Questions for chapter evaluation

Descriptive questions

1. What is accident disaster? List the type of accident disasters.
2. Define road accident. Write a note on road accident in India.
3. List the five top road accident prone cities and state and also describe the road accident prone highways of India.
4. What is rail accident? Write a note on rail network of India.
5. Write the causes of rail accidents in India. List the major rail accident in India.
6. What is air accident? List the major air accident in the world.
7. What is industrial accident? List some of the industrial accidents.
8. What are the fire safety rules? Write its prevention and precautions.
9. How bad weather affects the air transportation?

Objective questions

Fill in the blanks

i. The safest mode of transportation is ....

ii. The frequency of road accidents in India per hour is ....

iii. The road network of India is the .... largest network of the world.

iv. The rail network in India is .... km.

v. The air takes off from airport need clearance from ....

Mention true and false against each of the questions given below.

a. During the year 2019 there was no major rail accident in India

b. There was only one major air accident occurred in India since the year 2000 to 2019.

c. The death toll due to road accident is around 3 person per hour in India.

d. The Mumbai city of India has the largest road accident in India.

e. The major industrial accident in India occurred during 1994.

CHAPTER 6

POLLUTION

The material, which is harmful to human, plant kingdom or other habitats and have the capacity to alter the normal environmental conditions and also harmful impact on human health is known as pollution. The presence of any solid, liquid and gaseous materials in the atmosphere above the maximum concentration limit decided by Indian standard bureau and dangerous for living being is known as pollution. Generally, it is man-made wastes contaminate the environment is known as pollution or the pollution is the action of environmental contamination with man-made waste. This includes mainly land, water, and air. Pollution can come in various forms including the lesser-known noise, light, and thermal pollution.

Pollution is mainly of three types –Air, Water and Soil pollution.

6.1 Air Pollution

Any undesirable physical, chemical or biological change in air, which is able to adversely affect living and non-living beings is known as air pollution. Any contamination to the air by solid, liquid or gaseous wastes or by-products that can harmful for human health and welfare of plants and animals is known as air pollution. Air pollution is defined as an undesirable change in the physical, chemical or biological characteristics of air that may be harmful to human, other life, plants and cultural assets. In broad sense, pollution is the thermodynamic disorder that is the byproduct of energy conversion and the use of resources. Air pollution may be caused by natural as well as human activities. The pollutants are released by natural sources like volcanoes, coniferous forests, and hot springs. Human contributed a large portion of emissions of pollutants from industrial sources, power and heat generation, waste disposal, and the operation of internal combustion engines. Fuel combustion is the largest contributor to air pollutant emissions, caused by man, with stationary and mobile sources equally responsible.

6.1.1 Air Pollutant

Every day either, we come across about airborne chemicals in newspaper or see in the form of chemical disorder in air quality. All these are termed as air pollutants.

An air pollutant can be defined based on the concentration of chemical present in environment. The composition of clean air is used as a benchmark. If the concentration of a chemical is above the concentration of chemical present in air, it is then termed as an air pollutant.

There are two basic physical forms of air pollutants. The first is gaseous form. For example, sulfur dioxide, ozone and hydrocarbon vapours exist in the form of a gas. The gases lack definite volume and shape and the molecules are widely separated. The second form of air pollution is particulate matter, such as smoke, dust, fly ash and mists.

6.1.2 Common Air Pollutants

Let us now turn our attention to common air pollutants. There are six common pollutants:

i. sulfur di oxide ii. carbon monoxide, iii. nitrogen dioxide, iv. ozone, v. total suspended particulate matter, and vi. lead.

ii. The other class of air pollutants are hazardous and toxic air pollutants.

6.1.3 Types of Air Pollutants

Two main groups of air pollutants based on their way of emission.

i) Primary Pollutants

These emitted directly into the air are called primary pollutants, *e.g.*, particulates, sulfur dioxide, carbon monoxide, oxides of nitrogen and hydrocarbons.

ii) Secondary Pollutants

These are the pollutants produced through reactions between primary pollutants and normal atmospheric compounds, *e.g.*, ozone, photochemical products (aldehydes, PAN, PB_2N, olefin), photochemical smog (O_3, NO_x, hydrogen peroxides, organic peroxides) etc.

6.1.4 Sources of Air Pollution

The two major kinds of air pollution sources are stationary sources and mobile sources.

i) Stationary sources

It is point source either from an area or a region. These sources emit air pollutants from one or more controllable sites, from open areas exposed to wind processes and from well-defined areas within, which are several sources, *e.g.*, Industrial chimneys, municipal incinerators, commercial heating and open burning, small urban communities or areas of intense industrialization within urban complexes or agricultural areas sprayed with herbicides and pesticides.

ii) Mobile sources

These are emitters of air pollutants which move from place to place, while yielding emissions, *e.g.*, automobiles trucks, buses, aircrafts, trains etc.

6.1.5 Major air pollutants and their effects

There are five primary pollutants which together contribute more than 90 per cent of global air pollution. These are:

- Carbon monoxide, CO

- Oxides of Nitrogen, NOx
- Hydrocarbons, HC
- Sulphur di oxides, SOz
- Particulates

Carbon dioxide (CO_2): This is one of the air pollutants and increasing in the atmosphere at much greater rate after industrial revolution. The concentration of CO_2 in the atmosphere is 106 ppm in 1950 and now in 2019 its concentration is 400 ppm. It is estimated that approximately 9000–10000 billion ton CO_2 is deposited every year in the atmosphere by human activities. Though the human can able to tolerate very high concentration of CO_2 *i.e.,* upto 5000 ppm. The high concentration of CO_2 in atmosphere will be beneficial for C_3 plants because under higher concentration of CO_2 the photosynthesis rate will be increased., but the higher concentration may increase the temperature of the atmosphere and also the pH of the atmospheric water. It is estimated that 25 per cent increase in concentration of CO_2 will increase 0.1 percent pH of water.

Carbon monoxide: It is very hazardous gas for both human and plant. The major part of (90 per cent) the CO in atmosphere comes from the natural sources and remaining 10 per cent comes from incomplete burning of organic compounds, fires and automobiles. In presence of high concentration of CO, the leaves fall earlier from the plant and growths of roots on stem are some of the examples of the ill effect of CO on plant. The high concentration may develop heart, anaemia or respiratory disease in human. It may cause Hypoxia effect on human, which is harmful in breathing. It may also cause death on long exposure to high concentration.

Oxides of Nitrogen: The oxides of nitrogen (NOx) are mixture of gases that composed of nitrogen and oxygen. There are five oxides of nitrogen. NO_2 (Nitrogen dioxide), NO (Nitric oxide), N_2O (Nitrogenous oxide), NO_3 (Nitrogen trioxide) and N_2O_5 (Nitogen pentoxide) and among them only three nitrogen dioxide, nitric oxide , nitrous oxide are present in atmosphere and are very harmful and create pollution. It is emitted mainly in two forms: NO and NO_2. Nearly all NO_2 is emitted from automobiles and power plants that burn fossil fuels. The human seems more endangered by nitrous gases than plants. It irritates eyes, nose, throat and lungs and increase susceptibility to viral infections.

Hydrogen sulfide: It is highly toxic and corrosive gas. It is produced from natural sources such as geysers, swamps, marsh and petroleum refining and metal smelting. It causes health problem ranging from toxicity to death of humans and other animals. Hydrogen sulfide can cause severe poisoning and nerve damage in human and animals. Moreover, it can also affect plant enzymes and thus it produces irreversible damage.

Hydrogen fluoride: It is released from aluminium production, coal gasification and the burning of coal in power plants. Even at very low concentration (1ppb), may cause problems for plants and animals. Damage caused by hydrogen fluoride can be clearly recognized by the discolouration of the edges and tip of leaves, mostly brown in popular and black in birch leaves and leaves, curling up at the edges. Hydrogen fluoride emissions cause necroses at the apical part of fruits. They can result in deterioration in fruit flavour.

Chlorine and hydrogen chloride: Elementary chlorine and hydrogen chloride are used for the production of synthetic materials and insecticides. There is increased production of hydrogenchloride containing waste gases in the combustion of PVC containing plastic wastes. They cause severe damage to plants. The inhalation of these gases at higher concentrations leads to severe health damage. The mucous membrane of the respiratory tract is destroyed.

Ammonia: In the vicinity of intensive animal keeping and most sensitively to ammonia and alkylamine containing waste gases produced by decomposition of urea and uric acid or by the combustion of animal faeces. Ammonia often causes a dark brown to black colouring of the leaves of deciduous trees and potato leaves. Turnip show bright spots on young shoots or leaves.

Sulfur dioxide: The major anthropogenic source of Sulfur oxide (SO_2) is the burning of fossil fuels mostly coal in power plants. It is an irritant gas with adverse effects on the respiratory tract of humans and animals and also on the assimilation apparatus of plants. The disease like heart, dietary system (increase sugar content) and emphysema in human may be more under high concentration. Even after short-term influence, SO_2 concentration above 0.2 mg/m^3 of air can cause serious disorders in the assimilation plants organs. Building materials sensitive to acids such as limestones, sand stone, marble etc. are corroded and destroyed.

Hydrocarbons: The volatile organic compounds (VOCs) are composed of hydrogen and carbon, and may also contain elements, such as oxygen, nitrogen, sulfur, chlorine, and fluorine. The VOCs are used in the manufacture of glue and paints as solvents. Over 80 per cent of hydrocarbons, such as methane, butane and propane etc. are emitted through natural sources. The manmade sources of hydrocarbons include dry cleaning operations, auto paint shop, chemical plants, auto emissions, service stations and waste facilities. The adverse effects of hydrocarbons are numerous. At specific concentration, they are toxic to plants and animals or may be converted into harmful compounds through chemical changes in the atmosphere. Ethylene exposure causes chloroses and necroses, which were accompanied by the leaf edges curling up often followed by a wrapping of the leaf stem in tomato. In many plants, the youngest leaves react first. Further symptoms are growth depression and fading phenomena.

Ozone (O_3): It results from atmospheric interactions of oxygen and chlorine in presence of sunlight. High concentration of O_3 kills leaf tissue and even can kills whole plant. O_3 effect on animals including human, involves various kinds of damage, especially to eyes and the respiratory system. The consequences of O_3 exposure in plants become manifest as spot necroses. In tobacco plants, spot necroses appear on leaf surfaces and these symptoms are so called "Weather flecks" or ozone flecks. Tip necroses occur in onion leaves and other plants.

Chlorofluoro and Hydrofluorocarbon (CFCs and HFCs): Many times factories release greenhouse gases like carbon dioxide, chlorofluorocarbons (CFCs), methane, and nitrous oxide into the atmosphere. These products include Chlorine, fluorine, carbon and hydrogen. They are generated by cooling devices and during manufacturing of foam, plastic, propellants, cleaning materials, solvents etc. There life span in atmosphere is very high and they are dangerous for ozone and creates ozone hole. Plants yield is affected (lowered) and many disorders are observed in plants.

Particulate matter: Particulate matters are smoke, soot or dust, air born asbestos and small particles of heavy metals as arsenic, copper, lead and zinc, which are usually emitted from industrial facilities such as smelters. Particulate matter is particularly hazardous to the elders and those with respiratory problems such as asthma. Adverse effect of fly ashes is mainly seen in the pollution of vegetables and fodder plants. Fly ash sedimentation on fodder plants leads to a depreciation of feed-stuffs, reduced feed intake by animals, decreased milk production and some times physiological damage in pasture animals. The accumulation of lead, zinc and arsenic oxides in upper layers of soil leads to root depression and thus growth disturbances in plants. In extreme cases, the growth of cultivated plants can be completely stunted.

6.1.6 Toxic Air Pollutants

Toxic air pollutants may originate from natural sources such as volcanoes as well as from manmade sources such as stationary and mobile sources. The stationary sources serve as major contributors to air pollution since they include factories, refineries, or power pollutants, which are constantly emitting pollutants into the atmosphere.

Point sources include stationary facilities that emit sufficient amounts of pollutants to be worth listing. Area sources are all other point sources that individually emit small amount of pollutants. Mobile sources include automobiles, trucks, air-planes, ships, boats and lawnmowers. Natural sources are soil, water, vegetables, volcanic eruptions and lightning strikes.

Sources may also be classified as the way they generate emissions:

Transportation, Stationary combustion sources, Industrial processes, Solid waste disposal facilities and Miscellaneous.

i) Transportation sources

This category includes most emissions produced from transportation sources during the combustion process. The internal combustion engines fueled by gasoline and diesel are the biggest sources in this category. The other sources include trains, ships, farm tractors, planes, and construction machinery.

ii) Stationary combustion sources

In this category the sources only produce energy and the emission is a result of fuel combustion. The sources include power plants as well as home heating furnaces.

iii) Industrial processes

The sources, which emit pollutants during manufacturing of products are included in this category. Petrochemical plants, petrochemical refining, food and agriculture industries, chemical processing, metallurgical and mineral product factories and wood processing industries are the major sources of air emissions. The smaller sources include dry-cleaning, painting and degreasing operations.

iv) Solid waste disposal

This category includes facilities that dispose of unwanted trash. Refuse burning and open burning are important sources.

v) Miscellaneous

The sources, which do not fit in the above four categories are listed under this title. These sources include forest fires, house fires, agriculture burning, asphalt road paving and coal mining.

The pollutants typically released from the above sources are given in the Table 6.1.

Table 6.1: Source of typical air pollutants

Source	Typical air pollutants released
Transportation sources	carbon monoxide (CO), lead (Pb), nitrogen oxides (NOx), ozone (O_3)
Stationary sources	carbon monoxide (CO), lead (Pb), nitrogen oxides (NOx), particulate matter, sulfur dioxide (SO_2)
Industrial processes	carbon monoxide (CO), lead (Pb), nitrogen oxides (NOx), particulate matter, sulfur dioxide (SO_2)
Solid Waste Disposal	carbon monoxide (CO)
Miscellaneous (forest fires, coal mining etc.)	carbon monoxide (CO), nitrogen oxides (NOx), particulate matter, sulfur dioxide (SO_2)

6.1.7 Major Air Pollution Events

It is reported that the exposure to outdoor and indoor air pollution contributed to over 1.2 million deaths in India in 2017. The pollution is a global problem and many big cities of the world are polluted and the capital city of Delhi is the most polluted city of India for many years. Some major air polluted area of the world are listed below:

(i) A fog of three days occurred in Meuse valley (Belgium) in 1930 due to the air pollution (release of sulfur di oxide and other oxides) caused by steel power plant, sulphuric acid and zinc plants and resulted in death of 60 people, hundreds of people became ill and cattle became sick.

(ii) A fog of nine days occurred in Manchester, England in 1931, which had killed 592 people.

(iii) A severe smog caused by photochemical smoke and PAN (Peroxyacetylnitrate) occurred in 1945 at Los Angles (USA) and resulted in reduction in visibility, irritation to eyes and damage to vegetation.

(iv) There was a huge release of smoke due to the burning of coal and oil in Tokyo-Yokohama (Japan) in 1946. The asthma disease had spread and affected a large number of people.

(v) The 1948 plant emissions and atmospheric conditions in Donora, Penn. USA caused a 4-day fog and 7000 people were reported sick and 20 people died.

(vi) About 4000 people died and thousand hospitalized for respiratory troubles due to a polluted fog (smoke and SO2) on 5th December, 1952 at London (England).

(vii) Photochemical oxidant reacts with SO2, resulted in acid mist at Tokyo (Japan) in June, 1970 and about 6000 people suffered from eye irritation, sore throat and breathing difficulty.

(viii) In Bhopal (India), deadly methyl isocynate released due to failure of vent scrubber system on 3rd December, 1984 as a result of poised air pollution and about 2800 people died and one lac people severely affected by suffocation and cardiac failure.

(ix) There was a huge amount (55 ton) release of oleum gas from Shree Ram Food and Fertilizer Industries in Delhi on 4th December, 1985.

6.1.8 Control Measures of Air Pollution

Air pollution control may be defined as the various measures taken to meet certain emission standards. These measures may include changes in processes/raw materials or modification of equipment. Another method is the installation of devices at the end of process equipment to treat the exhaust gas stream. These devices are called air pollution control equipment.

i) Particulate control devices

- **Force field settlers:** These are equipments that use a field of force for the collection of particulate. There are three types of force fields: gravitational, centrifugal and electrical.
- **Fabric filters:** They are based on the principle of filtration for the removal of particulates.
- **Scrubbers:** They remove particulates from the exhaust gas stream by using water droplets for capturing them. Of the above, electrostatic precipitator and fabric filters possess the highest collection efficiencies.

ii) Some general control measures

1. Tall chimneys should be installed in industries to reduce air pollution on ground.

2. Better designed fuel burning equipments should be used in homes and industries, so that complete burning of fuel takes place.
3. Renewable and non-polluting sources of energy like solar energy and wind energy should be used, *e.g.*, solar cookers, wind machines etc.
4. Motor, vehicles should be maintained properly.
5. Strict emission control for automobiles, *e.g.*, Euro-II.
6. Zero pollutant automobiles should be manufactured and used, *e.g.*, electric cars.
7. There should be increased control on industrial activities and household activities that are known to contribute to air pollution.
8. More trees should be planted.

iii) Benefits of planting the tree

It has also been established that deforestation has contributed to the current global warming. Forests help in the reduction of carbon monoxide in the air. They improve human health by producing oxygen and improving the quality of air.

- Plants help to stop global warming by reducing the greenhouse gases
- Absorb more than a ton of harmful greenhouse gases over its lifetime (U.S. Environmental Protection Agency)
- One tree produces enough oxygen for four people every day (Tree Canada Foundation)
- Tree provides the equivalent cooling effect of ten room-size air conditioners operating 20 hours a day (U.S. Department of Agriculture)
- Tree provides an estimated $273 of environmental benefits in every year of its life (American Forests).

6.2 Water Pollution

Any undesirable materials or its mixture microbes, insecticide, pesticide, weedicide, industrial waste or domestic/sewer waste or sewage, when flown and mixed in water is known as water pollution. India was earlier known as the land of sacred rivers, but now it is gradually becoming a land of polluted and toxic water bodies. Water pollution is the action of environmental contamination with man-made waste into water. Water pollution is a serious problem in India as several of its surface water resources and groundwater reserves are contaminated by biological, toxic organic and inorganic pollutants. The source of this pollution could be a residue of pesticide, raw sewage, chemicals, trash, or fertilizer. Water pollution has severe human consequences. The groundwater is commonly containing less than 1mg/L nitrate. The nitrate limit for drinking water 10mg N/L and many pullulated groundwater contains 45mg N/L, which is far excess from this limit.

Water pollution can occur in oceans, rivers, lakes, and underground reservoirs and as different water sources flow together the pollution can spread.

6.2.1 Causes of Water Pollution

Water pollution has severe human consequences since less than 3 per cent of the Earth contains water that is potable or safe for drinking. The source of this waste could be raw sewage, chemicals, trash, or fertilizer. When a source for drinking water becomes contaminated with untreated sewage it is possible to spread diseases through the water like Giardiasis, a diarrheal disease caused by a parasitic protozoa, *Giardia lamblia*. These types of diseases can also be caused by bacteria, viruses, and other microorganisms. Another type of water pollution usually occurs, when water is contaminated with fertilizer. The water is then nitrified and phosphatized by the fertilizer chemicals. These added nutrients in the water can cause eutrophication or the process of changing a body water due to excessive plant nutrients. Some of the causes are:

- Heavy rainfall along with erosion
- Increased sediment from soil erosion
- Improper waste disposal and littering
- Disposal of industrial and domestic wastes into water reservoir
- Organic material decay in water supplies
- Eutrophication
- Deforestation and shifting cultivation
- Coal based Thermal power plant and mining

6.2.2 Eutrophication

It is process of receiving high concentration of nutrients, particularly nitrates and phosphates by any water bodies that stimulate excessive plant growth like algae, weeds etc.

The eutrophication promotes excessive growth of algae and aquatic plants, which then chokes the open-water areas, thus decreasing the amount of oxygen available in the water. Consequently, fish and other organisms may die off due to the lack of oxygen.

6.2.3 Chemical Demand of Oxygen (COD)

It is a measure of the capacity of water to consume oxygen during the decomposition of organic matter and the oxidation of inorganic chemicals such as Ammonia and nitrite. It is a measure of water and wastewater quality. The Chemical Oxygen Demand (COD) test is often used to monitor water treatment plant efficiency. A COD test aims to measure the amount of organic matter contained in a water sample. Higher COD levels mean a greater amount of oxidizable organic material in the sample, which will reduce dissolved oxygen (DO) levels. A reduction in DO can lead to anaerobic conditions, which is deleterious to higher aquatic life forms. Influent COD in normal domestic sewage is therefore generally 600–900 mg/l and it is then treated to at least 75–100 mg/l before discharge to minimise pollution potential. Raw sewage has COD 400 ppm. CPCB norms recommend that COD should be less than 250 ppm.

6.2.4 Biochemical Demand of Oxygen (BOD)

It is supposed to measure the amount of food (or organic carbons) that bacteria can oxidize. When BOD levels are high, dissolved oxygen (DO) levels decrease because the oxygen that is available in the water is being consumed by the bacteria. Since less dissolved oxygen is available in the water, fish and other aquatic organisms may not survive. BOD directly affects the amount of dissolved oxygen in rivers and streams. The greater the BOD, the more rapidly oxygen is depleted in the stream. This means less oxygen is available to higher forms of aquatic life. According to Central Pollution Control Board, limit of BOD for discharge of industrial and municipal waste waters into natural surface waters is less than 10 ppm *i.e.,* < 10 ppm.

A BOD level of 1–2 ppm is considered very good. There will not be much organic waste present in the water supply. A water supply with a BOD level of 3–5 ppm is considered moderately clean. Water with a BOD level of 6–9 ppm, the water is considered somewhat polluted because there is usually organic matter present and bacteria are decomposing this waste.

BOD up to 5 ppm is generally considered clean. Raw sewage has BOD 200 ppm. CPCB norms, recommend that BOD should be less than 30ppm.

6.2.5 Sources of Water Pollution

The sources, which cause water pollution can be classified into different categories:

i) Municipal and domestic oxygen demanding wastes, which contains decomposable organic matter and pathogenic agents. Municipal and domestic wastes include waste water from homes, commercial establishment, consist of domestic refuge, municipal garbage and other wastes like animal wastes, crops and yard wastes and garbages are mainly of organic origin.

ii) Most of the Indian rivers and freshwater streams are seriously polluted by industrial wastes or effluents, which come along waste water of different industries, such as petrochemical complexes, fertilizer factories, oil refineries, pulp and paper, textile, sugar and steel mills, tanneries, distilleries, coal washeries, synthetic material plants for drugs, fibers, rubber etc.

iii) Agricultural wastes include sediments, fertilizers, pesticides and farm animals wastes, which reach to water bodies through runoff and by leaching through soil to groundwater. Excessive use of agri-chemicals like fertilizers, pesticides. Even these pesticides have been detected in Arctic region.

iv) Mining wastes pollute streams and rivers. Ore fines and toxic substances carried by rain water into nearby water bodies, alter their chemistry and often makes the water unfit for human use.

v) Combustion of coal through thermal plants adds large amounts of pollutant (heavy metals) to the soil and water.

6.2.6 Water Pollutants and Their Effect

Industrial waste: The disposal of byproduct and other materials directly to water either in river or any other water reservoirs is major water pollutants. There are organic and inorganic materials in these pollutants like oil, plastic, phenol, metals, other chemical compounds. These pollutants decrease the oxygen in the water, which is harmful to water living organism. Fluoride is also mixed in water and is harmful and causes disease in human.

Domestic waste: The disposal of sewage, plastic along with other detergent materials into water is the second most important pollutant. These pollutants increase the growth of blue-green algae, bloom etc in the water.

Pesticide and insecticide: Hydrocarbons with chlorine or any chemical insecticides, weedicide, which is used for control of insect-pest are harmful for both man and animal.

Fertilizer: The uses of fertilizers in crops especially nitrogenous are dangerous to human and animal because they cause many diseases like cancer and increase the sugar content in the body.

Heavy metal: Mercury is very dangerous heavy metal and pollutes water and increases its toxicity, which ultimately affects human health at large scale. The infected fish, when used by human, likely to die. Similarly, cadmium, zinc and metal have affected human and animal health.

Petrol and oil: There are incidences, when large quantity of petrol and oil spread or release in sea due to some defect in vessels or accidents. They are harmful to many water organisms.

Power plant: Through combustion of coal based thermal power plant heavy metals like Fe, Zn, Cu, Mn, Pb, Cr, and Cd are entering into water bodies through rain water.

Mining: Mining wastes like ore, fluoride, arsenic, nitrate etc. are mixing in nearby water bodies and soil and pollute them.

6.2.7 Major Water Pollution Events

The largest source of water pollution in India is untreated sewage. Other sources of pollution include agricultural runoff and unregulated small-scale industry. Most rivers, lakes and surface water in India are polluted due to industries, untreated sewage and solid wastes. From the northern Himalayas to the sandy, palm-fringed beaches in the south, face acute water shortage, with close to 200,000 deaths every year from polluted water. The major water pollution events are listed below:

i) The "*itai-itai*" disease in Japan due to cadmium poisoning was traced due to discharge of waste water from a mine processing Cu, Zn in

'Jintsee River'. The river water is used in paddy crop irrigation. Similarly, Minamata was caused in 1955 by mercury poisoning.

ii) Lake Zurich in Switzerland and lake Erie in Canada are classic examples of induced eutrophication.

iii) Discharge of large quantities of oil along with the effluents from an oil refinery set the river aflame and resulted in suspension of water supply to the town and the refinery near Monghya, Bihar in 1968.

iv) Hoogly at Calcutta is receiving waste from power station, paper, jute, textile and chemical mills at an average rate of 52 tones/day. The water quality is worst than 4th grade as proposed by WHO.

v) The water of 'Yamuna' at Okhla industrial area for about 48 kms stretch is unfit even for irrigation purposes.

vi) Discharge of untreated waste water from a group of dye industries into the Kalu River near Mumbai, resulted in lowering of pH to 4.0.

6.2.8 Control Measures of Water Pollution

i) Physical treatment of water

Trickling filter– By this method all the organic material and undissolved matter may be filter out and remain in soil and water is collected in tank. Treating the water by water treatment plants before its use for any living being.

ii) Biological treatment

Biological agents may be used to decompose the plastic, oil and other petroleum products on or in the water.

iii) Implementation and adoption

- Direct disposal of industrial and domestic waste should be stopped.
- Ban on harmful pesticides.
- Use of recommended dose of pesticide and fertilizer in agriculture.
- Direct disposal of any by-product from manufacturing unit on or in the water should be banned.
- Plant more trees and they will reduce soil erosion and water pollution.

6.2.9 Industrial Waste Water Pollution

Industry is a huge source of water pollution, it produces pollutants that are extremely harmful to people and the environment. Many industrial facilities use freshwater to carry away waste from the plant and into rivers, lakes and oceans.

Pollutants from industrial sources include:

Asbestos – This pollutant is a serious health hazard and has carcinogenic effect. Asbestos fibres can be inhaled and cause illnesses, such as asbestosis, lung cancer, intestinal cancer and liver cancer.

Lead – This is a metallic element and can cause health and environmental problems. It is a non-biodegradable substance, so is hard to clean up once the environment is contaminated. It is harmful to the health of many animals, including humans, as it can inhibit the action of bodily enzymes.

Mercury – This is a metallic element and can cause soil health and environmental problems. It is a non-biodegradable substance, so is hard to clean up once the environment is contaminated. It is also harmful to animal health as it can cause illness through mercury poisoning.

Nitrates – The increased use of fertilizers means that nitrates are more often being washed from the soil and into rivers and lakes. This can be very problematic to marine environments.

Phosphates – The increased use of fertilizers means that phosphates are more often being washed from the soil and into rivers and lakes, which can be very problematic to marine environments.

Sulphur – This is a non-metallic substance that is harmful to marine life.

Oils – Oil does not dissolve in water, instead it forms a thick layer on the water surface. It is also harmful to fish and marine birds.

Petrochemicals – This is formed from gas or petrol and can be toxic to marine life.

6.2.10 Impact of Water Pollution on Plants

Photosynthesis is the process in which plants use the energy from sunlight to produce sugar, or glucose, by converting carbon dioxide into carbohydrates. However, when water is polluted, the capacity of water to dissolve gases such as carbon dioxide is negatively affected. Water pollution from substances can disrupt photosynthesis in plants. Plants that depend on photosynthesis for their survival, any interference in the photosynthetic process can affect the development and growth of plants.

Actually polluted water in the ground washes the essential nutrients that plants need from the soil. Water pollution makes the soil acidic and negatively affects the solubility of nutrient ions, such as iron, magnesium, potassium and calcium ions. As a result, water removes these nutrients more quickly from the soil and sends them into streams and lakes,. Calcium and magnesium, in particular, are critical for proper plant growth. Iron helps plants to create the pigment chlorophyll, which is necessary for food formation and potassium helps plants to use water. Without these nutrients, plants become more susceptible to drought, fungal infections and insects. Water pollution also leaves large amounts of aluminium in the soil, which can be harmful to plants.

6.3 Soil Pollution

Soil pollution is the action of environmental contamination with man-made waste on soil. It is possible that soil pollution may also contaminate the air and/or the water or vice versa. When the chemical waste may not be disposed of properly then the toxic substances may seep into the ground. These seeping chemicals can potentially contaminate a local body of water. It is also possible that these chemicals may give off a toxic vapour, thereby contributing to air pollution as well.

6.3.1 Causes of Soil Pollution

- Increased use of pesticide.
- Improper use of fertilizer.
- Polluted water used for irrigations.
- Damping-off industrial and domestic waste in or the soil.
- Spilling of materials by manufacturing sectors.
- Use of fly ash in agriculture.
- Mining.

Nitrates mixed with drainage water and deep into soil and thus may reach domestic wells, streams, lakes and flow groundwater. Thus creates problems to fish and other aquatic species. There are many sources of the inorganic chemical contaminants that can accumulate in soils.

Some of these are:

- Industrial organics-Pesticides (Insecticide, Fungicides, Herbicides, Nematocides etc.).
- Chlorinated hydrocarbon, organophosphate, organochlorine, carbamate, pyrethroid, hydroxyl acid, organic As and P acid pesticides.
- Carboxylic acid, Ammonosulfonyl acid, amide and anilide herbicides.
- Organometallic fungicides.
- Burning of fossil fuel.
- Smelting.
- Detergents.
- Fertilizers.
- Retardants.
- Cadmium, used in plating and manufacture of batteries.
- Copper.
- Manganese.
- Nickel.
- Zinc.
- Arsenic used in insecticide.

Source–Heavy metal in sewage sludge

Organic chemicals may enter the soil as contaminants in industrial and municipal organic wastes applied to or spilled on soils. Lubricants and fuel leaks, military explosives or as sprays applied to control pests. Pesticides are probably the most widespread organic pollutants associated with soils.

6.3.2 Control Measures of Soil Pollution

Physical and chemical method: Water-soluble and volatile chemicals may be removed by pushing or pulling air or water through the soil by vacuum extraction or leaching. The contaminations are either removed from the soil or sequestered in the soil.

Bioremediation: By use of enhanced plants or microbial action organic contamination is degraded into metabolic products. The polyacry aromatic hydrocarbons (PAHs) or pentachlorophenol(PCP) and trichloroethylene(TCE) can be broken down by soil bacteria.

Phytoremediation: Higher plants can be used in this process of soil treatments. In this plant, roots take up the pollutants from the soil. The plant may then either accumulate large amount of the contaminant in above ground biomass or it may metabolize the contaminant into harmless by-products.

Organic farming or Green farming: Cultivation of crops without the use of chemical fertilizer or pesticide is known as organic farming. Adaptation of organic and green farming will reduce the soil pollution too.

6.3.3 Land Pollution

Land pollution is the action of environmental contamination with man-made waste on land. Americans generate five pounds of solid waste every day, furthermore creating one ton of solid waste each year *(Kaufman and Franz)*. In an average day in the United States, people throw out 200,000 tons of edible food and throw 1 million bushels of litter out of their automobiles *(Kaufman and Franz)*. The main human contributor to pollution is landfills. Approximately half of our trash is disposed in landfills and only 2 percent of our waste is actually recycled.

Land pollution may also contaminate the air and/or the water or vice versa. Chemical waste may not be disposed of properly and toxic substances may seep into the ground.

6.3.4 Phyto-remediation

It is a process that uses various types of plants to remove, transfer, stabilize, and/or destroy contaminants in the soil and groundwater. It refers to the technologies that use living plants to clean up soil, air, and water contaminated with hazardous contaminants. In this process, green plants are used to treat and control wastes in water, soil, and air is an important part of the new field of ecological engineering. Through this technology plants remove the organic and inorganic wastes include metals and metalloids, some xenobiotic contaminants, and salts leachate, sewage, sludge, and other conventional wastes from soil and water (Figure 6.1).

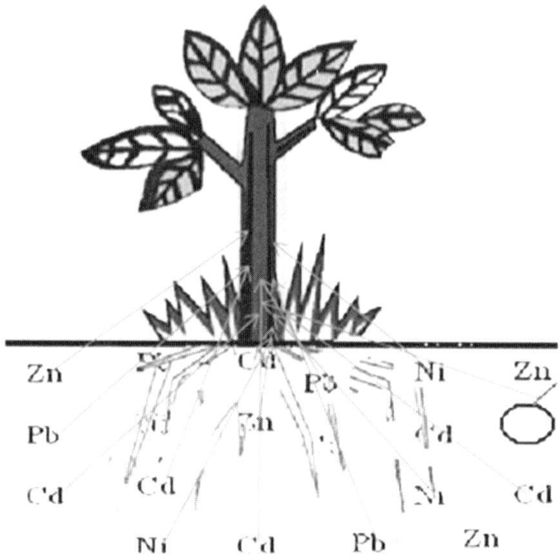

Figure 6.1 : Plant removing organic and inorganic wastes from soil

Questions for chapter evaluation

Descriptive questions

1. Define pollution. Write the type of pollution.
2. What is air pollution? Write a note on major air pollutants.
3. Describe the COD and BOD.
4. What are the control measures of air pollution. List the major events of air pollution of the world.
5. List some water pollution events of your region.
6. Define the soil pollution. How it is different from water pollution.
7. List the effect of soil pollution on crops.
8. What is the impact of air pollution on crops?
9. Define land pollution.
10. What is eutrophication? Why it is required?
11. How planting tree reduces the air pollution? List the major soil pollution events of India.

Objective questions

Fill in the blanks

i) The major air pollutant is ....

ii) Particulate matter is a type of ....pollutants.

iii) The BOD of safe water ....

iv) Water pollutants are classified in ....category.

v) The earth contains only .... per cent of safe drinking water.

Mention true and false against each of the questions given below

a) CFC is less harmful than CO_2.

b) HFC is a manmade air pollutant.

c) COD may be less for safe drinking water.

d) Use of pesticide does not increase soil pollution.

e) Organic farming or green farming is the control measure of soil pollution.

CHAPTER 7

DISASTER MANAGEMENT

Disaster can usually not be predicted, but having disaster management in place will help everyone be prepared, when disaster does strike. Being prepared will help with better recovery, staying safe and reducing injury as well as damage.

Manmade disasters are preventable. If people stick on rules that government has made for every purpose, such as by adopting the building bye laws, the building must be constructed accordingly. Similarly, by adopting the nuclear non-proliferation treaties, a country should not develop nuclear weapons for mass destruction. Stick to rules and conscientiously do keep watch on safety measures in industries then the accidents that cause suffering could be prevented. Natural Disasters too can be mitigated. By being prepared to face and respond to them effectively, we not only conserve the advances made by civilizations, but also minimize the losses. If no attention paid in this direction, it will difficult to deal with many challenges following a disaster. The situation may be warranted towards dealing with situations of injuries, property damage, hunger, thirsty, no electricity and no phones. Without prepared plan to deal such type of situations then all of these may cause problems. Therefore, disaster management helps put together a plan. It helps to know, what resources need and needs to gather. Now, there is personal family disaster management, but there is more extended disaster management that is needed as well. It includes having supplies that will be needed for distribution to the community, how these items will be distributed. It is also important to have a backup plan; such as if a distribution station has been harmed there should be an alternative.

In ancient India, the disaster management is also finding its space in the state responsibility domain. It was mentioned in *Kautilyas Arthashastra* that disaster management is a primary duty of the state. Every area from state and district to other countries has their own disaster management plan that needs to be created. There is no way to know how bad the disaster will be, so preparation is vital for survival.

Hence, Disaster management helps us to adapt ourselves to nature and learn to use it to our advantage. Taking care of our environment plays an important role in the mitigation of disasters. While economic development is necessary to match the demands of increasing population, we have to ensure that it does not result in environmental degradation. Development should, therefore, be planned in a judicious manner and in tune with sustaining and protecting our environment.

Disasters and their Management

Disasters are natural calamities and extreme events. India is considered as one of the most disaster prone regions with about 86 per cent area of the country prone to some types of disasters. Disaster can usually not be predicted or avoided, but disaster management will help everyone and make them to be prepared, when disaster does strike. Asian region is the region, which is most prone to natural disasters, facing more than 60 per cent of natural calamities, followed by America, Africa, Europe and Australia. Although disasters affect both the developed and developing countries, 90 per cent of the victims were from the developing nations. Every day there are mild earthquakes in one or the other part of the globe, but on an average about 20 large-scale disasters occur throughout the globe in one year.

7.1 Disaster Management

It is a systematic process based on the key management principles of planning, organizing, coordinating and controlling. The basic aim is to reduce the negative impact or consequences of adverse.

Disaster management includes various types of activities designed to mitigate the effects of disasters and emergency situations, and to provide a framework for helping people at risk to avoid or recover from the impact of the disaster. Managing disasters includes steps to be taken prior to, during, and after the disaster, and involve preparedness, mitigation, response and recovery.

Disaster management approach includes hazard and risk assessment, prevention, preparedness, prediction and warning, legislation, public education, relief and research. Disaster management is a continuous and integrated part of development process. Every country has its own disaster management plan. The new vision adopted by Indian government for disaster management comprises of the following concerns:

- Preparedness rather than crisis management
- Coordinated participatory approach rather than only state responsibility
- Technology up-gradation and deployment
- Information as a tool for disaster management
- Recognition of linkages between natural disasters and development
- Connecting specific programmes for managing natural disasters
- Emphasis on forecasting and warning using advanced technology
- Preparedness and mitigation through specific plan programmes

Common challenges for any disaster management plan are:

1. Inter-organizational coordination
2. Sharing information

3. Resource management
4. Advance warnings
5. Search and rescue
6. Use of mass media to deliver warnings to the public
7. Patient tracking
8. Hospital or healthcare agency damage
9. Volunteer management
10. Plan for organized improvisation

7.2 Disaster Management Components

The five important components of disaster management are:

1. **Prevention:** It is the action taken to eliminate/avoid natural hazards and their effects.
2. **Preparedness:** The set of activities and precautions that a community collectively takes before a disaster occurs, in order to reduce the impact of a disaster, and to cope with it efficiently.
3. **Mitigation:** It includes the measures taken to reduce both the effect of the hazard and vulnerable conditions in order to reduce the scale of a future disaster.
4. **Response:** It includes emergency management by various organizations to be taken in responding to disaster. Quick services are needed to be mobilized and functioning that is linked through national network.
5. **Recovery:** It is concerned with providing relief after the disaster. It deals with providing food and shelter to the disaster victims, restoring normal conditions and providing financial and technical assistance to rebuild.

Phases of Disaster

There are four phases of disasters:

i) Mitigation

Mitigation involves structural and non-structural measures taken to limit the impact of disasters. Structural mitigation is actions that change the characteristics of a building or its surrounding, examples include shelters, window shutters, clearing forest around the house. Non-structural mitigation on personal level mainly takes the form of insurance or simply moving house to a safer area.

Personal mitigation is mainly about knowing and avoiding unnecessary risks. This includes an assessment of possible risks to personal/family health and to personal property. In floods prone areas houses can be built on poles/stilts, as in much of southern Asia are more risky.

Prevention: It is concerned with policies and programs to prevent the recurrence of natural disasters and covers the long-term aspect of such disasters.

It can involve such actions as:

- Laws and regulations restricting deforestation to prevent mudslides
- Relocating or elevating structures to minimize the effects of floods
- Securing shelves and water heaters to walls in earthquake zones
- Developing, adopting and enforcing building codes and standards
- Engineering roads and bridges to withstand earthquakes
- Using fire-retardant materials in new construction to reduce the risk of fire

ii) Preparedness

Preparedness measures can take many forms including the construction of shelters, implementation of an emergency communication system, installation of warning devices, creation of back-up life-line services (*e.g.*, power, water, sewage), and practising evacuation plans.

Personal preparedness focuses on preparing equipment and procedures for use when a disaster occurs, *i.e.*, planning. Simple measures can help and prepare the individual for sitting out the event or evacuating, as necessary.

The essential elements include:

- Planning
- Training personnel
- Tabletop disaster drills

iii) Response

The response phase of an emergency may commence with search and rescue, but in all cases the focus will quickly turn to fulfilling the basic humanitarian needs of the affected population. This assistance may be provided by national or international agencies and organizations. Effective coordination of disaster assistance is often crucial.

Response activities may include:

- search and rescue
- triage
- acute medical care
- fire fighting
- sheltering victims
- relocating medical records

iv) Recovery

The recovery phase starts after the immediate threat to human life has subsided. During reconstruction, it is recommended to consider the location or construction material of the property.

Recovery is defined as the actions taken to return to normal following an event.

- Repairing buildings
- Replacing holmes

Disaster management is fundamentally disaster risk management. Total of all activities, programmes and measures, which can be taken up before, during and after a disaster with the purpose to avoid a disaster, reduce its impact or recover from its losses is called Disaster Risk Management. There are three stages of the disaster risk management, which is collectively called Disaster Management Cycle. Broadly, there are six phases in Disaster Management Cycle *viz.* Prevention, mitigation, preparedness, response, recovery and reconstruction.

The Prevention, Mitigation and Preparedness include Pre-disaster activities focused on reducing the human and property losses caused by a potential hazard; Response, Recovery and Reconstruction include the Post-disaster initiatives (Figure 7.1) taken in response to a disaster with a purpose to achieve early recovery and rehabilitation of affected victims and communities.

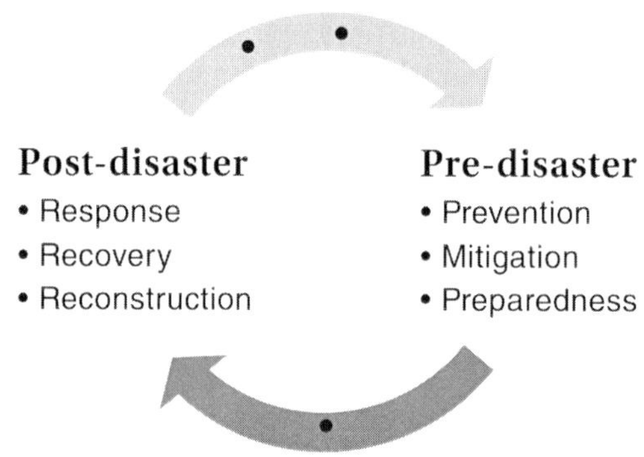

Figure 7.1 : Pre and post disaster phase

7.3 Disaster Management Cycle (DMC)

Disaster risk management includes sum of all activities, programmes and measures, which can be taken up before, during and after a disaster with the purpose to avoid a disaster, reduce its impact or recover from its losses. The three key stages of activities that are taken up within disaster risk management are:

1. **Before a disaster (pre-disaster).**

   All the activities, which are taken to reduce human and property losses *e.g.,* carrying out awareness campaigns, strengthening the existing weak structures, preparation of the disaster management plans at household and community level etc. Measures taken under this stage are termed as mitigation and preparedness activities.

2. **During a disaster (disaster occurrence)**

   All initiatives are taken to ensure to meet the needs of people and minimize their suffering. Activities taken under this stage are called emergency and response activities.

3. **After a disaster (post-disaster)**

   All initiatives are taken with a purpose to achieve an early recovery and rehabilitation of affected communities, these actions are called as response and recovery activities. The conceptual diagram for planning disaster management is shown in Figure 7.2.

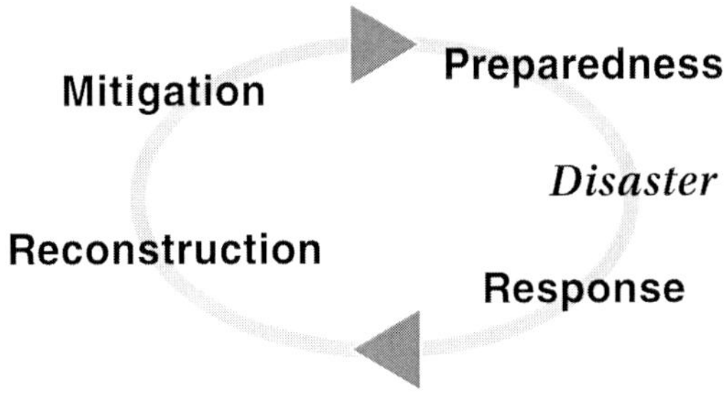

Figure 7.2 : Conceptual diagram for planning of disaster management

Through disaster management, we cannot completely counteract the damage, but it is possible to minimize the risks through early warning, provide developmental plans for recovery from the disaster, generate communication and medical resources, and aid in rehabilitation and post-disaster reconstruction. The exchange of correct information following the event is important, in order to ensure the resources necessary to support response and recovery activities. The seventy two hours after occurrence of hazardous event is the most crucial time for coordination among relief organizations.

7.4 Disaster Mitigation

Disaster mitigation is the ongoing effort to lessen the impacts of disasters have on people and property. Fewer people and communities would be affected by natural disasters with the use of this process. Due to the varying degree of each natural disaster, there are different mitigation strategies for each.

7.4.1 Mitigation Measures

All measures are taken to reduce the loss of life, livelihood and property caused by natural disaster, either by reducing vulnerability or by modifying, where possible, the hazard.

- Mitigation elements
- Natural hazard (H)
- Vulnerability (V)
- Elements at risk (E)
- Specific risk (Rs)
- Risk (Rt)

In recent years remote sensing and GIS technology have been effectively used for disaster monitoring and mitigation. The Scientific and Technical Committee (STC) of International Decade for Natural Disaster Reduction (IDNDR) identified the following actions to be undertaken as part of disaster reduction strategy:

- Identification of hazard zones and hazard assessment.

- Vulnerability and risk assessment.
- Awareness at the level of decision and policy makers.
- Monitoring, prediction and warning.
- Early-intervention measures.
- Short-term protective measures and preparedness.
- Long-term prevention measures, including non-structural and structural measures.

7.4.2 Disaster Risk Reduction: This includes event and loss minimization through quick recovery plans. This may be achieved by reducing vulnerability and increasing coping capacities.

7.4.3 Functions of Disaster Reduction: To understand the functions of disaster reduction, its elements are shown in Figure 7.3. The following functions /works may be undertaken to reduce the disaster impact.

- Hazard risk analysis
- Multi-hazard impact analysis

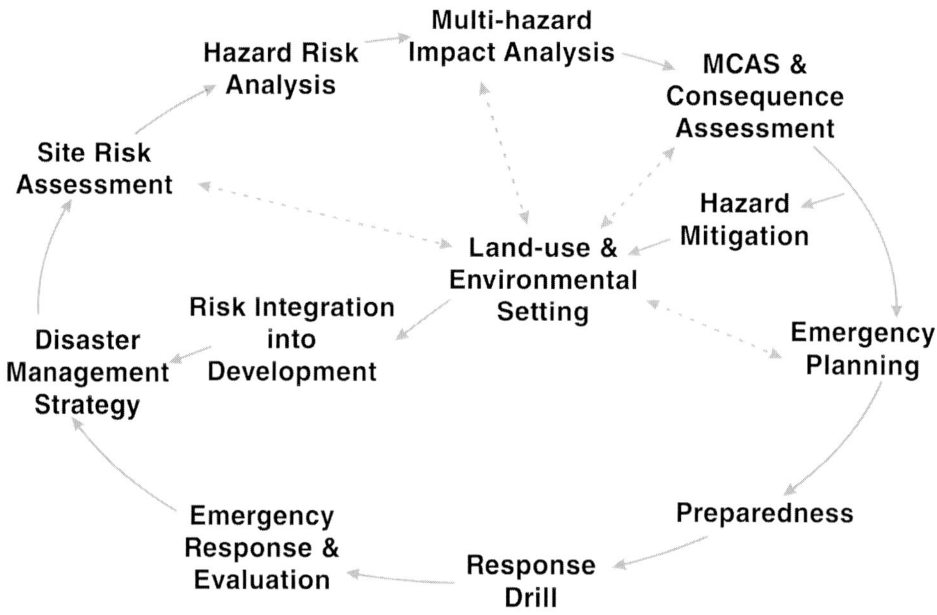

Figure 7.3 : The function of disaster management.
(*Source:* NDMA, Department)

7.4.4 Disaster Mitigation Plan

Natural disaster mitigation is essential for sustainable development because the effects of disasters pose heavy strains on development efforts and divert funds from other needed purposes. A mitigation plan is often divided into various sectors such as infrastructure, health, communications, livelihoods, agriculture, etc. For each sector, mitigation measures are listed, and identification of responsible or 'nodal' persons for disaster management activities. This would also include the contingency plan and emergency communication systems. The longer-term mitigation efforts would include improving building codes, safe locations for new residential areas, emergency communication systems, cross-training of potential responders.

i) Efforts to mitigate natural disasters at national level:

India is affected by different types of natural disasters from time immortal. Unchecked population growth and poorly constructed buildings/structures are aggravated the effected of natural disasters. Disasters like tsunamis and earthquakes, which have been the most destructive, along with the floods and droughts that arise from extreme weather conditions, are expected to get worse due to adverse impact of climate change. The country has a long experience of management of disaster and its mitigation, but population pressure, unplanned urban centres, lack of disaster preparedness and climate change pose serious threats in rural and highly populated cities. Urbanization in India is increasing the risks of disasters, which, in turn, are exposing the vulnerability to various hazards.

In recent past country experienced the Bhuj earthquake (2010, the tsunami (2004); the Kashmir earthquake (2005); Mumbai heavy rainfall (2006); the Kosi disaster (2008); the Leh cloud burst (2010) and recently the Sikkim earthquake (2011). All these disasters were managed and responded by coordination between various ministries of Indian Government.

Disaster management in India has evolved from an activity-based setup to an institutionalized structure; from single faculty domain to a multi-stakeholder setup; and from a relief-based approach to a multi-dimensional approach for reducing risk. The beginnings of an institutional structure for disaster management came during British period following the series of disasters that hit the country. These were the Famines of 1900, 1905, 1907,1943, and the Bihar-Nepal Earthquake of 1937. Over the past century, the structure for managing disasters in India has undergone substantive changes in its composition, nature and policy. The British administration chose a reactive approach for managing disasters and set up relief departments which provided emergency relief during disasters. This was an activity-based setup which was functional only in the post-disaster scenarios.

Post-independence, the task for managing disasters continued to rest with the Relief Commissioners in each state and functioned under the Central Relief Commissioner, with their, role limited to a delegation of relief material and money in the affected area. Frequent occurrence of floods and droughts in the country further limited the scope of disaster management. Every five-year plan addressed flood disasters under "Irrigation, Command Area Development and Flood Control". Till this stage, the disaster management structure was activity-based, functioning under the relief departments, therefore a single-faculty domain setup, and receiving weight only in terms of 'financing relief'.

The emergence of a permanent and institutionalized setup began in the decade of 1990s. The disaster management cell was established under the Ministry of Agriculture, following the declaration of the decade of 1990 as the International Decade for Natural Disaster Reduction (IDNDR) by the UN General Assembly. Further, India witnessed series of disasters, such as Latur Earthquake (1993), Malpa Landslide (1994), Odisha Super Cyclone (1999) and Bhuj Earthquake (2002) which reoriented the policy action and led to the shift from financing to overall management.

With frequent occurrence of disasters, there is an increasing consensus amongst the states for setting up an effective disaster management mechanism at the state level. The administrative responsibility for disaster preparedness and management, and the task of responding to any kind of disasters has been entrusted to the states. The centre intervenes only, when the magnitude of disaster escalates beyond the capability of the state authorities. However, the responsibility for coordinating disaster response and relief operation is that of the Union Ministry of Home Affairs (MHA).

The Union Ministry of Home Affairs is the nodal ministry for disaster management. The National Crisis Management Committee (NCMC), headed by the Cabinet Secretary and constituted under the Ministry, oversees the various disaster related activities in the country. It functions as a decision-making body and gives directions to the Crisis Management Group (CMG). The CMG is the vital body that actually deals with all the matters related to relief activities in the case of any major disasters. Further, for the execution of policies formulated in the ministry, the Central Relief Commissioner (CRC) is designated as the nodal officer for coordination of relief operation. The office of the CRC receives all the information related to forecasts and warnings from the Indian metrological department and issues directions for an action plan in response to the situation. Various other ministries are also assigned the responsibility for hazard identification and risk.

7.4.5 Disaster Mitigation Initiatives at National Level

i) First Country workshop on Disaster Management

The first country workshop on disaster management was one step in the efforts of the natural disaster mitigation. This workshop was held in July 1993 at New Delhi and conducted by Indian Institute of Public Administration in collaboration with Asian Disaster preparedness centre, Bangkok and Ministry of Agriculture.

ii) Establishment of National Institute for Disaster Management (NIDM)

On the recommendation of the first country workshop on disaster management and during the ESCAP seminar; Ministry of Agriculture recognized the Indian Institute of Public Administration as the National Centre for Disaster Management (NCDM). Later on, this centre is converted by the Indian government into a separate Institute and renamed as National Institute of Disaster Management in the year 2002.

The first systemic attempt of natural disaster relief measures could be traced after Odisha Famine of 1886 and the first famine commission was appointed by then Government of India under the chairmanship of the late Sir George Campbell, which laid the foundation of the famine policies. The second step was appointment of a Famine commission in 1980 following the famines of 1873 and 1876–78, which formulated the famine code for administration of relief and suggested two acts Land Improvement Loan act of 1883 and Agriculture Loan act of 1884.

It would also be of interest to note that following the great famine of 1876–78, a special fund of Rs. 15 million, known as the famine relief and insurance Fund, was set apart every year from 1882. Half of it was earmarked for development of railways and irrigation if it was not spent on famine relief.

Famine commission 1900 had recommended that in all future settlements, any increase of assets due to construction of well, tanks or other artificial sources of irrigation should be permanently exempted from assessment to revenue.

The U.N. General Assembly has declared the nineties (1990–2000 A.D.) as the International Decade for Natural Disaster Reduction (IDNDR) for minimizing the adverse impact of the natural calamities and to formulate national disaster mitigation programmes, as well as economic, land-use and insurance policies for disasters prevention, and, particularly in the developing countries, to integrate them fully into their National Development Programmes. Indian efforts in the IDNDR is quite significant in the setup of national level institute of disaster management at Delhi know as NIDM. NIDM function as a nodal institution at national level for training and research in disaster management.

The Disaster Management Act enacted on December 23, 2005 empowers the setting up of an effective disaster management system that extends across the whole of India. The Disaster Management Policy of India was framed in 2009 on the basis of this Act. Now, the handling disaster situations has been changed from relief-centric to a holistic, multi-dimensional, and multi-disciplinary approach involving diverse scientific, engineering, social, and financial processes. It encompasses the entire scope of disaster management activities, *i.e.,* prevention, mitigation, preparedness, response, relief, and rehabilitation.

iii) Efforts to mitigate natural disasters at international/global level

The International Strategy for Disaster Reduction (ISDR) is a strategic framework adopted by United Nations Member States in 2000. The ISDR guides and coordinates the efforts of a wide range of partners to achieve a substantive reduction in disaster losses. It aims to build resilient nations and communities as an essential condition for sustainable development.

The UN General Assembly recognized the need for global action to reduce the occurrence of natural disaster and their impact. The resolution No. 44/236 of UN General Assembly has drawn attention to the implications of natural disasters for the economic development of any country and proposed following actions:

- Scientific technical input in the management of natural disasters
- Community involvement

- Awareness programmes and
- Building disaster mitigation as an integral component of development programmes.

Disasters are complex problems and generally arising from the interaction between the environment and the development. It requires complex responses drawing on a wide range of skills and capacities. Thus it needs the cooperation between multilateral development agencies, national and local governments, non-governmental organizations, natural and social scientists, technical specialists and the vulnerable communities. There are many global meeting, efforts and conferences (United Nations Conferences in Stockholm in 1972, Rio de Janeiro in 1992, and the World Summit on Sustainable Development held in Johannesburg), towards sustainable development like Paris agreement. Such declaration calls for the concerted efforts of all UN member states to reducing the occurrence and impact of natural disasters.

The international decade for disaster reduction (1990–1999) the UN General Assembly in 2000 founded the ISDR (International Strategy for Disaster Reduction), a coalition of governments, UN agencies, regional organizations and civil society organizations. In 2002, the UN published a document entitled Living with Risk: A Global review of disaster reduction initiatives. In 2005, a major reform within the UN system resulted in UN and UNDP, becoming increasingly concerned about actively engaging in enhancing disaster risk programmes at the country level. The road map towards the implementation of the United Nations Millennium Declaration touches on areas which are closely linked to vulnerability to natural hazards such as ensuring environmental stability, the eradication of extreme poverty and hunger, and promoting gender equality.

iv) International Day for Disaster Reduction (IDDR)

The International Day for Disaster Reduction was designated the second Wednesday of every October to increase awareness of disaster reduction through IDNDR activities. In 2001, the UN General Assembly agreed to maintain observance of IDDR as a vehicle to promote a global culture of natural disaster reduction, including disaster prevention, mitigation and preparedness.

v) Disaster Mitigation Activities

Recent years have witnessed a series of catastrophes. Agencies of the United Nations, international and regional organizations, and many governments have acted quickly to dispatch rescue teams to implement relief and provided manpower, material and financial assistance to the affected areas in various forms. Major participants are:

International strategy for disaster reduction: In response to recognized need for better preparedness and response strategies towards natural disaster mitigations, the United Nations University (UNU), together with the World Seismic Safety Initiative (WSSI), the Inter-national Centre for Disaster Mitigation Engineering (INCEDE) of the University of Tokyo, and Stanford University have established a Global Network on National Disaster Reduction (UNU,1994). The objective of this network is to initiate, implement and encourage the development of programmes to manage and reduce the risk of property and environmental damage and human suffering, which are associated with natural disasters.

Losses due to natural disasters cannot be nullified, but they may be mitigated by integrating new and existing knowledge, and by managing risk through various structural and non-structural strategies. International cooperation is needed to meet the challenge of this complex problem. The Global Network in Natural Disaster Risk Management (GLO-DISNET) is intended to act as a catalyst in formulating effective programmes to deal with the issues and global risk associated with natural disasters. GLO-DISNET will focus primarily on initiating research and training programmes in developing countries. The network will incorporate a country-specific approach to identify a key vulnerable hazardous area in each concerned country and apply the successful experience of one country to another. The approach taken is multi-hazard and multi-faceted, ranging from disaster preparedness through response strategies to community recovery.

The Network involves the following activities:

- Develop a general framework for global research and training programme involving multidisciplinary natural disaster management.
- Bring together related research on natural disaster reduction under a single international network.
- Increase interaction among interested researchers for promoting related research and training, reduce research overlap, and promote a comparative approach.
- Create a global database of research findings about natural disaster reduction for dissemination to interested UN bodies and organizations.
- Provide reliable, detailed information for effective environmental management, risk management and government strategies to effectively work to reduce the risks of natural disasters.

The United Nations General Assembly launched the decade 1990–2000 as International Decade for Natural Disaster Reduction (IDNDR, 1990–2000) in 1989. The goal of IDNDR is to formulate national disaster-mitigation programmes, as well as economic, land use and insurance policies for disaster prevention, to integrate them fully into their national development programme. The decade would enable governments to focus on hazard vulnerability and risk assessment, disaster prevention, sustainable development, effective early warning systems, sharing of knowledge and transfer of technology. It emphasized on concerted international action, particularly in developing countries to handle the loss of life, property damage, social and economic disruption caused by natural disaster. The IDNDR secretariat located in Geneva is a part of the United Nations Department of Humanitarian Affairs.

7.4.6 Disaster Mitigation Initiatives at International Level

International Association of Emergency Managers

The International Association of Emergency Managers (IAEM) is a non-profit educational organization dedicated to promoting the goals of saving lives and protecting property during emergencies and disasters. The mission of IAEM is to serve its members by providing information, networking and professional opportunities, and to advance the emergency management profession.

International Recovery Platform

The International Recovery Platform (IRP) was conceived at the World Conference on Disaster Reduction (WCDR) in Kobe, Hyogo, Japan in January 2005. As a thematic platform of the International Strategy for Disaster Reduction (ISDR) system, IRP is a key pillar for the implementation of the Hyogo Framework for Action (HFA) 2005–2015: Building the Resilience of Nations and Communities to Disasters, a global plan for disaster risk reduction for the decade adopted by 168 governments at the WCDR. The key role of IRP is to identify gaps and constraints experienced in post-disaster recovery and to serve as a catalyst for the development of tools, resources, and capacity for resilient recovery. IRP aims to be an international source of knowledge on good recovery practice.

Red Cross/Red Crescent societies

National Red Cross/societies often have pivotal roles in responding to emergencies. Additionally, the International Federation of Red Cross and Red Crescent Societies (IFRC) may deploy assessment teams, *e.g.,* Field Assessment and Coordination Team (FACT) to the affected country, if requested by the National Red Cross or Red Crescent Society. After having assessed the needs Emergency Response Units (ERUs) may be deployed to the affected country or region. They are specialized in the response component of the emergency management framework.

United Nations

Within the United Nations system, responsibility for emergency response rests with the Resident Coordinator within the affected country. However, in practice international response will be coordinated, if requested by the affected country's government, by the UN Office for the Coordination of Humanitarian Affairs (UN-OCHA), and then they deployed a UN Disaster Assessment and Coordination (UNDAC) team.

World Bank

Since 1980, the World Bank has approved more than 500 operations related to disaster management, amounting to more than US$40 billion. These include post-disaster reconstruction projects, as well as projects with components aimed at preventing and mitigating disaster impacts, in countries such as Argentina, Bangladesh, Colombia, Haiti, India, Mexico, Turkey and Vietnam to name only a few.

Common areas of focus for prevention and mitigation projects include forest fire prevention measures, such as early warning measures and education campaigns to discourage farmers from slash and burn agriculture that ignites forest fires; early-warning systems for hurricanes; flood prevention mechanisms, ranging from shore protection and terracing in rural areas to adaptation of production; and earthquake-prone construction.

In a joint venture with Columbia University, the World Bank has established a Global Risk Analysis of Natural Disaster Hotspots.

In June 2006, the World Bank established the Global Facility for Disaster Reduction and Recovery (GFDRR), a longer, term partnership with other aid donors to reduce disaster losses by mainstreaming disaster risk reduction in development, in support of the Hyogo Framework of Action. The facility helps developing countries for fund development projects and programs that enhance local capacities for disaster prevention and emergency preparedness.

European Union (EU)

Since 2001, the EU adopted the Community Mechanism for Civil Protection, which started to play a significant role on the global scale. The main role is to facilitate co-operation in civil protection assistance interventions in the event of major emergencies, which may require urgent response actions. This also applies to situations, where there may be an imminent threat of such major emergencies.

The major part of the community mechanism is to act as a Monitoring and Information Centre (MIC). It is part of Directorate-General for Humanitarian Aid and Civil Protection of the European Commission and accessible 24 hours a day. It gives countries access to a platform, to a one-stop-shop of civil protection. Any country inside or outside the Union affected by a major disaster can make an appeal for assistance through the MIC. It acts as a communication hub at headquarters level between participating states, the affected country and dispatched field experts. It also provides useful and updated information on the actual status of an ongoing emergency.

Yokohama Strategy and Plan of Action for safer world

World Conference on Natural Disaster Reduction Guidelines for Natural Disaster prevention, preparedness and mitigation was held at Yokohama, Japan, during 23–27 May, 1994. The conference adopted Yokohama strategy and related plan of action for the rest of decade and beyond.

International Strategy for Disaster Reduction (ISDR)

ISDR was set-up at the end of International Decade for National Disaster Reduction, during the 1990s, to carry forward the mission "building disaster resilient communities by promoting increased awareness of the importance of disaster reduction as an integral component of sustainable development with the goal of reducing human, social, economic and environmental losses due to the natural hazards and related technological and environmental disasters".

Inter-Agency Task Force (IATF)

An Inter-Agency Task Force comprising representatives of 16 agencies and organizations. It is programs of the regional entities, civil society and professional organizations to provide the policy guidance for the International Strategy for disaster reduction.

Hyogo Framework for Action(HFA)

The world conference on Disaster Reduction was held at Kobe, Japan during 18–22 January, 2005. The conference had adopted the Hyogo Framework for Action (HFA) 2005–15: Building the Resilience of Nations and Communities to disasters.

HFA mandated the International Strategy for Disaster Reduction (ISDR) to develop "generic, realistic and measurable indications" for assessing the progress in the implementation of the framework, keeping in mind available resources of individual states.

The United Nations International Strategy for Disaster Reduction secretariat (UNISDR) works with governments, organizations and civil society to further the disaster risk reduction agenda through implementation of the Hyogo Framework for Action, a ten-year plan of action adopted in 2005 by 168 governments to protect lives and livelihoods against disasters, and provides credible evidence and advocacy for increased investments in disaster risk reduction and climate change adaptation.

> ***Coordinate*** *in the international efforts in disaster risk reduction,* ***Advocate*** *for greater investment in disaster risk reduction actions,* ***Campaign*** *to build global awareness of disaster risk reduction and to* ***Inform*** *and connect people by providing services and practical tools.*
>
> ***Connect the countries to implement the programmes for*** *disaster risk reduction,* ***Promote*** *the effective reduction of climate risks,* ***Encourage*** *the establishment of regional and national platforms for disaster risk reduction, and* ***Provide*** *guidance for the implementation of the Hyogo Framework for Action risk and progress in disaster risk reduction*

7.5 Disaster Contingency Plan

The International Strategy for Disaster reduction observed that the number of storms, droughts, and floods have increased threefold over the past 30 years. Natural disasters can happen anywhere, anytime, with little or without warning. India has recurring natural disasters, seasonal in their patterns, just like the floods in the monsoon season. Under these situations, there is a need to have a contingency plan devised for such circumstances. Contingency planning for natural disasters call for specific strategies and actions to deal with particular problems like drought, floods, landslides, fires, and typhoons. Contingency Plans include a monitoring process and phases for activating planned actions.

Contingency planning has to move with the times, never static, always updated to accommodate new or increased risks.

Local authorities may have option to choose a strategic partner with demonstrated emergency preparedness experience. Apart from the available civil society organizations the Red Cross and Red Crescent Societies have excellent resources for creating effective disaster contingency plans.

7.6 Statutory Provision

As per the Government of India factory and manufacturer guidelines, every factory should prepare on-site emergency plan for disaster control measures for his factory. Manufacture, storage and import of hazardous chemicals should also prepare and keep up to date on-site emergency plan containing details how major accidents will be dealt with on the site. Every factory and manufacturer, stoker involve in industrial activity should inform/display the name of the person, who is responsible for safety on the site and names of those, who are authorized to take action in accordance with the plan in case of an emergency.

7.7 Emergency Planning

This is the details of planning on-site emergency plan to control disaster in every factory. Major accidents may cause emergency and it may lead to disaster, which may cause heavy damage to plant, property, harm to person and create adverse effects on production. An emergency situation arises all of a sudden and creates havoc and damage to person, property, production and environment. Therefore such situations and risks should be thought in advance and it should be planned before in hand to tackle them immediately and control them within the shortest time. On-site Emergency Plan and detailed disaster control measures for every factory should be known to the workers employed therein and to the general public living in the vicinity of the factory, so that the safety measures required to be taken in the event of an accident taking place.

7.7.1 Emergency

Each major hazardous factory should prepare an emergency plan incorporating details of action to be taken in case of any major accident/disaster occurring inside the factory. The plan should cover all types of major accident/occurrences and identify the risk involved in the plant. Mock drills on the plan should be carried out periodically to make the plan foolproof and persons are made fully prepared to fight against any incident in the plant. The plan will vary according to the type of industry and emergency.

Many disasters like Bhopal gas tragedy, Chernobyl nuclear disaster etc. have occurred in many places in the world causing heavy loss of life and property.

It is defined as an accident/incident that has potential to cause serious injuries or loss of life. It may cause extensive damage to property, serious disruption both in production and working of factory and may adversely affect the environment. The following factors may cause major emergency.

(i) Plant failure.
(ii) Human error.
(iii) Vehicle crash.
(iv) Sabotage.
(v) Earthquake.
(vi) Natural Calamities.

I. Types of Emergency

Emergency is of two types, on-site emergency and off-site emergency

On-site Emergency

It is a situation, when the accident/incident takes place within the factory and its effects are confined to the factory premises, involving only the persons working in the factory and the property inside the factory, it is called as on-site emergency.

On-site Emergency Plan: The on-site emergency plan is prepared and shall be documented in a printed form insufficient copies to give all concerned for knowledge, study and easy follow-up. The emergency plan shall be rehearsed and practised at regular intervals to test efficiency of personnel, equipments, coordinated efforts and to increase confidence and experience to operate such plan. It may contain:

1. Site plan and topographic plan.
2. Plan showing the fire fighting facilities.
3. Plan showing hazardous material storage area.
4. Material safety data sheets for hazardous chemicals.
5. Facilities available in main control centre.
6. List of emergency equipment.
7. List of Safety Equipment.
8. List of important telephone numbers and addresses.
    i. Nearest hospitals and ambulance service centre.
    ii. Nearest fire station.
    iii. Govt. Officials.
    iv. Transport provider.
9. Names and address and contact telephone number of Key Personnel.

Off-site Emergency

It is a situation, when the accident is such that it affects inside the factory are uncontrollable and it may spread outside the factory premises, it is called as off-site emergency.

Off-site Emergency Plan: The main objective of the plan are:

i. To save lives and injuries.

ii. To prevent or reduce property losses and

iii. To provide for a quick resumption of normal situation or operation.

7.7.2 Emergency Action Plan

The action plan should consist designated emergency control, centre/room, etc.

Main elements of on-site emergency plans:

- Leadership and Administration.
- Role and Responsibilities of Key Personnel.
- Emergency action.
- Light and Power.
- Source of energy control.
- Protective and rescue equipment.
- Communication.
- Medical care.
- Mutual Aid.
- Public relation.
- Protection of vital records.
- Training.
- Periodical revision of plan.

7.7.3 Emergency Control Centre (ECC)

This is the main centre where the operations to handle the emergency are directed and coordinated. Maximum facilities to be made available in the emergency control are :

i. Internal and external communication.

ii. Computer and other essential records.

iii. Daily attendance of workmen employed in factory.

iv. Storage of hazardous material records and manufacturing records.

v. Pollution records.

1) Storage area of hazardous materials:

a. Store the hazardous materials in a safe place.

b. Storage of safety equipments.

c. Fire fighting system and additional source of water.

d. Site entrance, roadway and emergency exist.

e. Assembly points.

f. Truck parking area.

g. Surrounding location.

2) Assembly Points

It is space or a safe or pre-determined point or place far away from the plant, wherein case of any emergency all the workers should be assembled. In case of emergency personnel evacuated from the affected areas are to be assembled at the assembly point. The plant workers, contract workers and visitors should assemble in assembly point in case of emergency.

3) Alarm System

Alarm system varies and will depend on the size of the works area. Simple fire bell, hand-operated siren – break open type, fire alarm etc. Automatic alarm may be needed for highly hazardous nature of plant.

Siren for Emergency: Siren for an emergency should be different from the normal siren. The emergency siren should be audible to a distance of 5 KM radius. The emergency siren should be used only in case of emergency.

4) Escape Route

The escape route from each plant should be clearly marked. The escape route is the shortest route to reach out of the plant area to open area, which leads to assembly point. This route should be indicated on the layout plan attached to the on-site emergency plan.

5) Evacuation

All non-essential staff should be evacuated from the emergency site. As soon as the emergency siren rings the workers have to shut down the plant and move to the assembly point. The plant shut down procedure in case of emergency should be prepared and kept ready and responsible person should be nominated for the purpose.

6) Counting of Personnel

All personnel working in the plant should be counted. Time office person should collect the details of personnel arriving at the assembly point. These should be checked with the attendances of regular workers, contract workers present in the site on the day of emergency. The accident control should be informed and arrangement should be made for searching missing person in the emergency affected area. The employees' address, contact number of next to kin should be maintained in the time office, so that during emergency relatives of those affected due to emergency may be informed accordingly.

7) Mutual Aid System

Mutual aid scheme should be introduced among industries, so that in case of emergency necessary help from mutual aid partner may be extended. Essential elements of this scheme are Mutual aid must be a written document signed by the Chief Executive of the industries concerned.

- Specify key personnel who are authorized to give requisition of materials from other industries.

- Specify the available quantity of material/equipment that can be spared.
- Mode of requisition during emergency.
- Mode of payment/replacement of material given during an emergency.
- May be updated from time to time based on experience gained.

Mock drills

Mock drill on emergency planning should be conducted once in 6 months and sequence of events should be recorded for improvement of the exercise. Exercises on On-site Emergency Planning should be monitored by the high officials of the organization and the plan is reviewed every year.

Emergency facilities

The following facilities should be provided in any factory to tackle any emergency at any time.

1. Fire protection and fire fighting facilities.
2. Emergency lighting and standby power.
3. Emergency equipment and rescue equipment.
4. Safety Equipment.

7.8 Communication System

Communication is a key component to control an emergency. The following communication systems may be provided in the plant:

- Walky-Talky.
- Telephone (internal and external).
- Cell phone.
- Intercom/paging.
- Runners (verbal or written messages).

7.8.1 Disaster Communications

Reliable communications are necessary during any emergency or natural disaster, such as flood, fire, earthquake, hurricane, or even the unthinkable act of terrorism. In case of a natural disaster or emergency situation, normal communications will be disrupted and calling for immediate help will be very much needed. Though almost all have a cell phone, but using them will be difficult immediately after a disastrous event. Previous emergency situations have shown that cellular systems are not designed to handle that high volume of calls and will quickly overload, getting through to anyone is nearly impossible. The cell phone network as the result of damaged Tower Antennas or Base Stations, your cell phones could be inoperable for days or even weeks.

A land-line is a better way to immediately call during or after a disaster, but disconnected during many disaster situations. The most reliable way to communicate or call for help during or after a disaster is to use a 2–meter Amateur

Band Ham Radio. Anyone can use them, they can be inexpensive, and can be the size of a standard walkie-talkie to easily pack away for use in emergencies. Emergency Responders use the 2–meter Amateur Radio Band. The 2–meter Amateur Band frequencies are not locally limited or affected in times of emergencies, Ham radio operators are often the first contacted to assist in major disasters with communications between the public and emergency services, such as law enforcement, fire response, and emergency medical services. The American Red Cross has recognized that the 2–meter Amateur Band is the most effective way to provide communication and Ham Radio operators provide 90 percent of the coordination efforts during a major emergency. If a major disaster occurs, having these radios could save lives.

7.9 Risk Assessment

Risk assessment is most essential before preparing any off-site emergency plan. Hazardous factories and their hazard identification, other hazard prone areas, specific risks, transportation risk, storage risks, pollution risks by air and water pollution, catastrophic risks such as disasters, natural calamities, acts of God, earthquake, landslide, storm, high wind, cyclone, flood, scarcity, heavy rain, lightening, massive infection, heavy fire, heavy explosion, volcano, heavy spill, toxic exposure, environmental deterioration etc., risks from social disturbances, risks from the past accidents must be considered, while carrying out risk assessment for a particular area (district) from which the offsite emergency plan is to be prepared.

7.9.1 Central Control Committee (CCC)

As the offsite plan is to be prepared by the Government, a Central Control Committee shall be formed under the Chairmanship of the District Collector. Other officers from Police, Fire Service, Factory Inspectorate, Medical Department shall be incorporated as members of the Central Control Committee. Under the Central Control Committee the following committees shall be constituted under the control of the District Collector.

i. Incident and Environment Control Committee.
ii. Fire Control Committee.
iii. Traffic control, Law and order, Evacuation and Rehabilitation Committee.
iv. Medical help, Ambulance and Hospital Committee.
v. Welfare, Restoration and Resumption Committee.
vi. Utility and Engineering Services Committee.
vii. Press, Publicity and Public Relations Committee.

The off-site emergency plan shall be prepared by the District Collector in consultation with the factory management and Govt. agencies. The plan contains up to date details of outside emergency services and resources such as Fire Services, Hospitals, Police etc. with telephone number. The district authorities are to be included in the plan area.

a. Police Department.
b. Revenue Department.
c. Fire Brigade.
d. Medical Department.
e. Municipality.
f. Gram/Village Panchayat.
g. Railway Department.
h. Telephone Department.
i. Factory Department.
j. Electricity Department.
k. Pollution Control Department.
l. Explosive Department.
m. Press and Media.

Mock exercises on Off-site plan should be carried out at least once in a year to train the employees, up to date the plan, observe and rectify deficiencies.

7.10 Disaster Managements

Every type of disaster needs some special type of actions for its management. A few major disaster managements are listed below:

7.10.1 Earthquake Management

Community preparedness: Community preparedness is vital for mitigating earthquake impact. The most effective way to save you even in the slightest shaking is 'DROP, COVER and HOLD'.

Planning: The Bureau of Indian Standards has published building codes and guidelines for safe construction of buildings against earthquakes. Before the buildings are constructed the building plans have to be checked by the Municipality, according to the laid down bylaws. Many existing life-line buildings, such as hospitals, schools and fire stations may not be built with earthquake safety measures. Their earthquake safety needs to be upgraded by retrofitting techniques.

Engineered structures: Buildings need to be designed and constructed as per the building by-laws to withstand ground shaking. Architectural and engineering inputs need to be put together to improve building design and construction practices. The soil type needs to be analyzed before construction. Building structures on soft soil should be avoided. Buildings on soft soil are more likely to get damaged even if the magnitude of the earthquake is not strong. Similar problems persist in the buildings constructed on the river banks, which have alluvial soil.

Public education: Educate the public on the causes and characteristics of an Effect of Soil type on ground shaking. It can be created through sensitization and training programme for community, architects, engineers, builders, masons, teachers, government functionaries and students.

7.10.2 Tsunami Management

It is not possible to prevent a tsunami, in certain tsunami-prone countries, some measures have been taken to reduce the damage caused onshore. Japan has implemented an extensive programme of building tsunami walls of up to 4.5m (13.5 ft) high in front of populated coastal areas. Other localities have built flood gates and channels to redirect the water from incoming tsunamis. For instance, the tsunami which hit the island of Hokkaido on July 12, 1993 created waves as much as 30m (100 ft) tall, as high as a 10-story building. The port town of Aonae on Hokkaido was completely surrounded by a tsunami wall, but the waves washed right over the wall and destroyed all the wood-framed structures in the area. The wall may have succeeded in slowing down and moderating the height of the tsunami, but it did not prevent major destruction and loss of life. Some other systematic measures to protect coastlines against tsunamis may be undertaken to minimize losses, which includes:

(a) Site Planning and Land Management: Within the broader framework of a comprehensive plan, site planning determines the location, configuration, and density of development on particular sites and therefore, an important tool in reducing tsunami risk or natural hazard areas is recommended as the first land-use planning strategy. This strategy is designed to keep development at a minimum in hazard areas. In areas, where it is not feasible to restrict land to open-space uses (parks and recreation), other land use planning measures can be used. These include strategically controlling the type of development and uses are not allowed in hazard areas, and avoiding high-value and high occupancy uses to the greatest degree possible.

(b) Engineering structures: Most of the habitation of the fishing community is seen in the coastal areas. The houses constructed by them are mainly of light weight materials without any engineering inputs. Therefore there is an urgent need to educate the community about the good construction practices that they should adopt such as:

Site selection– Avoid building or living in buildings within several hundred feet of the coastline as these areas are more likely to experience damage from tsunamis.

Elevate coastal homes– Most tsunami waves are less than 3 meters in height. Construct the structure on a higher ground level with respect to mean sea level. Elevating house will help reduce damage to property from most tsunamis.

- Construction of water breakers to reduce the velocity of waves.
- Use of water and corrosion-resistant materials for construction.
- Construction of community halls at higher locations, which can act as shelters at the time of a disaster.

(c) Flood Management: Flooding will result from a tsunami. Tsunami waves will flood the coastal areas. Flood mitigation measures could be incorporated.

7.10.3 Cyclone Management

Coastal belt plantation — Green belt plantation along the coastal line in a scientific interweaving pattern can reduce the effect of the hazard. Providing a cover through green belt sustains less damage. Forests act as a wide buffer zone against strong winds and flash floods. Without the forest the cyclone travel freely inland. The lack of protective forest cover allows water to inundate large areas and cause destruction. With the loss of the forest cover each consecutive cyclone can penetrate further inland.

Hazard mapping — Meteorological records of the wind speed and the directions give the probability of the winds in the region. Past records and paths can give the pattern of occurrence for particular wind speeds. A hazard map will illustrate the areas vulnerable to cyclone in any given year. It will be useful to estimate the severity of the cyclone and various damage intensities in the region. The map is prepared with data inputs of past climatological records, history of wind speed, frequency of flooding etc.

A shelter with special feature to withstand cyclones and floods. Traditional homes can be improved by building in disaster resistant features. Such homes could withstand cyclones with moderate speeds.

Policies should be in place to regulate land use and building codes should be enforced.

Engineered structures — Structures need to be built to withstand wind forces. Good site selection is also important. Majority of the buildings in coastal areas are built with locally available materials and have no engineering inputs. Good construction practice should be adopted with good site selection. It is advised to construct on stilts or earth mound. Houses can be strengthened to resist wind and flood damage. All elements holding the structures need to be properly anchored to resist the uplift or flying off of the objects.

- Buildings should be wind and water resistant.
- Protect river embankments. Communication lines should be installed underground.
- Provide strong halls for community shelter in vulnerable locations.

7.10.4 Flood Management

Flood Management In India, was done by a systematic planning for flood management commenced with the Five Year Plans, particularly with the launching of National Programme of Flood Management in 1954. During the past many years, different methods of flood protection structural as well as nonstructural have been adopted in different states depending upon the nature of the problem and local conditions. Structural measures include storage reservoirs, flood embankments, drainage channels, anti-erosion works, channel improvement works, detention basins etc. and non-structural measures include flood forecasting, flood plain zoning, flood proofing, disaster preparedness etc. The flood management measures undertaken, so far have provided reasonable degree of protection to an area of 15.81 million hectares throughout the country.

Flood mitigation measures could be incorporated by adopting following actions:

Improving vegetation cover: The roots of the plants and trees keep the soil intact and prevent erosion and slow runoff. Coastal belt plantation prevent or lessen flooding. The use of tree planted in rows will act as a windbreak. Coastal shelterbelt plantations can be developed to break severe wind speeds. It minimizes devastating effects. The Odisha calamity has also highlighted the need for urgent measures like shelterbelt plantation along with cyclone-prone coastal areas. Species chosen for this purpose should not only be able to withstand the impact of strong cyclonic winds and also check soil erosion.

Mapping of the flood prone areas: It is a primary step involved in reducing the risk of the region. Historical records give the indication of the flood inundation areas and the period of occurrence and the extent of the coverage. The warning can be issued looking into the earlier marked heights of the water levels in case of potential threat. In the coastal areas, the tide levels and the land characteristics will determine the submergence areas. Flood hazard mapping will give the proper indication of water flow during floods.

Land use control: This will reduce danger of life and property when water inundate the floodplains and the coastal areas. The number of casualties is related to the population in the area at risk. In areas, where people already have built their settlements, measures should be taken to relocate to better sites to reduce vulnerability. No major development should be permitted in the areas which are subjected to high flooding. Important facilities like hospitals, schools should be built in safe areas. In urban areas, water holding areas can be created like ponds, lakes or low-lying areas.

Construction of engineered structures in the flood plains and strengthening of structures to withstand flood forces and seepage. The buildings should be constructed on an elevated area. If necessary to build on stilts or platform.

Flood Control aims to reduce flood damage. This can be done by decreasing the amount of runoff with the help of reforestation, protection of vegetation, clearing of debris from streams and other water holding areas, conservation of ponds and lakes etc. Flood Diversion include levees, embankments, dams and channel improvement. Dams can store water and can release water at a manageable rate., but failure of dams in earthquakes and operation of releasing the water can cause floods in the lower areas. Flood Proofing reduces the risk of damage. Measures include the use of sandbags to keep flood water away, blocking or sealing of doors and windows of houses etc.

7.10.5 Drought management

There are various mitigation strategies to cope up with drought.

- **Short term measures:** This includes awareness of the availability of safe drinking water, water conservation techniques, agricultural drought management strategies like crop contingency plans, construction of rain water harvesting structure. Awareness can be generated by the print, electronic and folk media.
- **Drought Monitoring:** It is continuous observation of the rainfall situation, availability of water in the reservoirs, lakes, rivers etc. and comparing with the existing water needs in various sectors of the society.
- **Choice of crop and contingency planning:** If probability of drought is anticipated more area should be brought under short duration and low water requirement crops like moth, cowpea, guar, moong, urd etc. through rainwater harvesting in houses and farmers' fields increases the content of water available. Adopt contingency planning as recommended by local administration.
- **Water Harvesting:** Water harvesting should be carried out in all watershed area and other areas by allowing the runoff water in farm ponds, ponds. In situ soil conservation by contour bunds, contour cultivation raised bed planting etc. helps increase water availability for crop production.
- **Ground water utilization:** Ground water use should be rationalized for human and animal use. Ban on major building construction, supply of water for household work and animal should be given priority than other activities, otherwise it will be misused.

Long term planning

- **Improving the Monitoring and forecasting of drought:** There is need to improve the technique for an early warnings for drought. Network of rainfall and other meteorological data should be analyzed at weekly intervals and may be provided to each block for better monitoring of drought. Forecasting of drought may be issued in the first week of June month, so that contingency planning may be fine-tuned and adopted by framers.
- **Establishment of perennial vegetation cover:** It is necessary to maintain vegetation cover on the soil to prevent soil erosion and also to reduce the evaporation rate. There is a need to adopt agro-horticulture, silvopasture and agro-pasture models depending on the area of drought.
- **Livelihood planning** identifies those livelihoods, which are least affected by the drought. Some of such livelihoods include increased off-farm employment opportunities, collection of non-timber forest produce from the community forests, raising goats, carpentry etc.

- **Drought planning:** The basic goal of drought planning is to improve the effectiveness of preparedness and response efforts by enhancing monitoring, mitigation and response measures.

Planning for establishing drought taskforce, which is a team of specialists who can advise the government in deciding to deal with drought situation, establishing a coordination mechanism among various agencies, which deal with the droughts, providing crop insurance schemes to the farmers to cope with the drought related crop losses, and public awareness.

7.11 Landslide Management

Hazard mapping locates areas prone to slope failures. This will help to avoid building settlements in such areas. These maps will also serve as a tool for mitigation planning.

Land use practices such as:

- Areas covered by degraded natural vegetation in upper slopes are to be afforested with suitable species. Existing patches of natural vegetation (forest and natural grass land) in good condition, should be preserved.
- Any developmental activity initiated in the area should be taken up only after a detailed study of the region has been carried out.
- In construction of roads, irrigation canals etc. proper care is to be taken to avoid blockage of natural drainage.
- Total avoidance of settlement in the risk zone should be made mandatory.
- No construction of buildings in areas beyond a certain degree of slope.

- **Retaining Walls** can be built to stop land from slipping. These are constructed to prevent smaller size and secondary landslides that often occur along the toe portion of the larger landslides.
- **Surface Drainage Control Works:** The surface drainage control works are implemented to control the movement of landslides accompanied by infiltration of rain water and spring flows.
- **Engineered structures** with strong foundations can withstand or take the ground movement forces. Underground installations (pipes, cables, etc.) should be made flexible to move in order to withstand forces caused by the landslide.
- **Increasing vegetation cover** is the cheapest and most effective way of arresting landslides. This helps to bind the top layer of the soil with layers below while preventing excessive run-off and soil erosion.
- **Insurance** will assist individuals, whose homes are likely to be damaged by landslides or by any other natural hazards.

7.12 Action Plan for Biological Disaster Management in India

Biological disaster could arise from a source located either inside the country or outside the country. Management of such a situation could be dealt effectively only. if there is a disaster plan well integrated into the system and also there is mechanism of post-disaster evaluation. An early warning system framework in India may be adopted for monitoring evaluation and prediction of the biological system.

The inter disaster stage is a crucial period in which pre-disaster planning in terms of system development should be done. It needs an action plan of following elements:

(a) Constitution of a Crisis Management Structure

- Identification of Nodal Officers for Crisis Management at District, State and Central Level.
- Identification of Focal points for control of epidemic at District, State and Central Level.
- Constitution of advisory committees – Administrative and Technical
- Preparation of contingency plan including Standing Operating Procedure at District, State and Central Level.

(b) System of Surveillance

- System of information collection at District, State and Central Level.
- System of data analysis
- System for flow of information from District to State and to Central level during crisis period.
- Establishment of control rooms at District, State and Central Level.

(c) System of Epidemiological Investigation

- System of field investigation.
- System of active surveillance.
- Arrangement for support facilities.

(d) Confirmation of pathogens by laboratory set up

- System of laboratory investigation at District, State and Central Level.
- Quality Control of Laboratory Practices.

(e) Training to different level workers

7.12.1 Early Warning Signal and Detection

Early identification of an outbreak of disease of national and international public health importance shall require knowledge of early warning signals amongst all the echelons of health care providers. Some of the suggested early warning signals, which must command quick investigation by professionals may include followings:

- Sudden high mortality or morbidity following acute infection with short incubation period
- The disease area should be mapped and excluded in endemic areas
- Occurrence of cases, which are difficult to diagnose with available clinical and laboratory support and they are non-responsive to conventional therapies
- Clustering of cases/deaths in time and space with high case fatality rate
- Unusual clinical or laboratory presentations
- A comprehensive lists of all the trigger events that shall attract immediate attention of local public health machinery need to be developed by a group of experts.
- By suspicion: Management Plan should aim to identify crisis situation at a very early stage. This can be done only by suspecting danger of impending disaster by local health employees (at village by village health guide, at sub-centre level by multipurpose worker)
- Alertness of institution dealing with emergency health and available medical services.

Question for chapter evaluation

Descriptive questions

1. Write a note on disaster management. What are the new visions adopted by Government of India for disaster management?
2. Discuss the disaster management components and its different phases.
3. Describe the international level initiatives of disaster mitigation.
4. What is emergency plan? Write the types of emergency plans.
5. Write the disaster management for earthquake and Tsunami.
6. Write in brief the biological disaster management.
7. What is flood and drought management strategy in India?
8. Write the landslide management points.
9. Write about cyclone management points in India.

Objective questions

Fill in the blanks

(i) Disaster management cycle consists of .... stages.

(ii) International days for disaster reduction is .... Wednesday of every October.

(iii) The first national level workshop in India was held at New Delhi in .... 1993.

(iv) .... is a resolution of UN general assembly for attention of implication of natural disaster for economic development of any country.

(v) The international decade for natural disaster reduction was ....

Mention true and false against each of the questions given below.

(a) Response is concerned with providing relief after disaster.

(b) National institute of disaster management was renamed in 2002 from national centre for disaster management.

(c) Mitigation is a measure taken to reduce the impact of hazard and vulnerability.

(d) Harvesting is a important point for flood management?

(e) Structural engineering is a point for earthquake, cyclone and Tsunami management.

CHAPTER 8

DISASTER MANAGEMENT IN INDIA

In India, the National Disaster Framework is a roadmap, which covers institutional mechanic; disaster prevention strategy, early warning system, disaster mitigation, preparedness, response and human resource development.

The expected inputs, areas of intervention and agencies to be in at the national, state and district levels have been identified and listed in the roadmap. This roadmap has been shared with all the state Governments and Union Territory Administrations. Ministries, Departments of Government of India and the State Governments/UT Administrations have been advised to develop their respective roadmaps taking the national roadmap as a broad guideline. There is a common strategy foundation for the action being taken by the entire participating organization and stakeholders.

The approach is being put into effect through:

(a) Institutional changes
(b) Enunciation of policy
(c) Legal and techno-legal framework
(d) Mainstreaming Mitigation into Development process
(e) Funding mechanism
(f) Specific schemes addressing mitigation
(g) Preparedness measures
(h) Capacity building
(i) Human Resource Development and, above all, community participation.

The administrative responsibility for disaster preparedness and management and the task of responding to any kind of disasters has been entrusted to the states. The centre intervenes, only when the magnitude of disaster escalates beyond the capability of the state authorities.

8.1 Disaster Management Act

The Disaster Management Act enacted on 23rd December, 2005 empowers the setting up of an effective disaster management system that extends across the country (details of the act is given in Annexure II). The Disaster Management Act, 2005 came into the statute book on 26th December, 2005 by a Gazette notification.

The Act provides a legal and institutional framework for "the effective management of disasters and for matters connected therewith or incidental thereto." The orientation for handling disaster situations was also changed from relief-centric to a holistic, multi-dimensional, and multi-disciplinary approach involving diverse scientific, engineering, social, and financial processes. It encompasses the entire scope of disaster management activities, *i.e.,* prevention, mitigation, preparedness, response, relief, and rehabilitation with adequate financial and administrative powers and creation of the National Institute of Disaster Management (NIDM) with the mandate of undertaking training and capacity building, develop training modules on various aspects of disaster management, undertake research and documentation, formulate and implement comprehensive human resource development plan covering all aspects of disaster management. Provide assistance in national level policy formulation and provide assistance to state governments and state training Institutions. Further, the Disaster Management Act of 2005 provides for the constitution of the following institutions at national, state and district levels:

i) The National Disaster Management Authority (NDMA) at the centre chaired by the Prime Minister is responsible for laying down the national policies, plans and guidelines for disaster management.

ii) The State Disaster Management Authorities (SDMA) for formulating policies and plans for disaster management in the states.

iii) District Disaster Management Authorities (DDMA) for planning, coordinating, and establishing systems for disaster management at the district level in accordance with the guidelines laid down by the national and state authorities.

The act also provides guidelines for the creation of National Disaster Response Fund (NDRF), National Mitigation Fund (NMF), establishment of funds by state Government and allocation of funds by ministries and departments for emergency procurement.

Concurrently, the National Institute of Disaster Management (NIDM) at New Delhi was established after the upgradation of the National Centre for Disaster Management (NCDM). It focuses on human resource development, capacity building, training, research, documentation, and policy advocacy in the field of disaster management.

Ministry of Agriculture set up the National Centre for Disaster Management (NCDM) at IIPA, New Delhi in 1995.

Ministry of Home Affairs upgraded NCDM in a National Institute is known as National Institute of Disaster Management (NIDM) in October, 2003 and situated in New Delhi.

8.2 National Policy on Disaster Management

The Government of India has established institutional and policy mechanisms for the response, relief, and rehabilitation. The Disaster Management Policy of India was framed in 2009, based on national disaster management act. The policy laid down the roadmap/directions for all government endeavours. India's National Policy on Disaster Management was approved by the union cabinet of India on 22nd October, 2009 with the aim to minimize the losses to lives, livelihoods and property, with a vision to develop a holistic, proactive, integrated, multi-disaster oriented and technology driven strategy with an emphasis on building strategic partnerships at various levels.

The Policy is also intended to promote a culture of prevention, preparedness and resilience at all levels through knowledge, innovation and education. It encourages mitigation measures based on environmental sustainability. It seeks to mainstream disaster management into the developmental planning process and provides for institutional and financial arrangements at national, state, and district-levels for disaster prevention, mitigation, preparedness and response. It also ensures adequate budgeting for disaster mitigation activities in all ministries and departments. The policy includes community based disaster management, capacity development in all spheres, consolidation of past initiatives and best practices and cooperation with agencies at national and international levels with multi-sectorial synergy.

> *The national policy on disaster management was approved by the Government in November 2009. The objectives of the policy document lay down on every aspect of holistic management of disasters in the country.*

Objective of the disaster management policy

- A holistic and pro-active approach for mitigation and preparedness will be adopted.
- Each ministry/Department of the central/state government to ensure that disaster reduction elements are integrated with development planning.
- Each ministry/Department to prepare and execute a disaster management plan for its sector role.
- Ensuring mitigation measures based on state-of the art technology and environmental sustainability.
- Putting in place a streamlined institutional technological framework in order to create and preserve the integrity of an enabling regulatory environment and a compliance regime.
- Promoting a productive partnership with the media to create awareness and contributing towards capacity development.
- Ensuring efficient response and relief with a caring approach towards the needs of the vulnerable sections of the society.

- ➢ Development of technological framework for enforcement of Indian standards and codes relevant to multi hazard resistant designs and review and revised the control regulations and relief codes and manuals.
- ➢ Establishment of national centres for human resource development and research.
- ➢ Involvement of community, NGO and other organization in disaster management.

The policy is reviewed by national disaster management authority, the apex body responsible for disaster management In India.

Salient Features of Disaster Management Policy:

- *Preparedness rather than Crisis Management*
- *Coordinated participatory approach rather than only state responsibility*
- *Technology up-gradation and deployment*
- *Information as a tool for disaster management*
- *Recognition of linkages between natural disasters and development*
- *Connecting specific programmes for managing natural disasters*
- *Emphasis on forecasting and warning using advanced technology*
- *Preparedness and Mitigation through specific Plan Programmes*
- *Disaster management as a continuous and integrated part of development process*

8.3 Disaster Management in India

The Disaster Management Act 2005 has provided the legal and institutional framework for disaster management in India at the national, state and district levels. In the federal polity of India the primary responsibility of disaster management vests with the State Governments. The Central Government lays down policies and guidelines and provides technical, financial and logistic support while the district administration carries out most of the operations in collaboration with central and state level agencies.

The nodal agency for disaster management in India is Ministry of Home Affairs (MoHA), New Delhi. This ministry is assisted by high power committees on disaster management. The other two important departments are National Disaster Management Authority and National Institute for Disaster Management.

The Cabinet Committee on Management of Natural Calamities (CCMNC) oversees all aspects relating to the management of natural calamities including assessment of the situation and identification of measures and programmes considered necessary to reduce its impact, monitor and suggest long term measures for prevention of such calamities, formulate and recommend programmes for public awareness for building up society's resilience among them. The Cabinet Committee on Security (CCS) deals with the matters relating to nuclear, biological and chemical emergencies. The National Crisis Management Committee (NCMC) under the Cabinet Secretary oversees the Command, Control and Coordination of the disaster response.

The National Disaster Management Authority (NDMA) under the Chairmanship of the Prime Minister is the apex body responsible for laying down policies, plans and guidelines for disaster management and for coordinating their enforcement and implementation throughout the country. The policies and guidelines will assist the Central Ministries, State Governments and district administration to formulate their respective plans and programmes. NDMA has the power to approve the National Plans and the Plans of the respective Ministries and Departments of Government of India. The general superintendence, direction and control of National Disaster Response Force (NDRF) are vested in and will be exercised by the NDMA.

The National Executive Committee (NEC) is mandated to assist the NDMA in the discharge of its functions and further ensure compliance of the directions issued by the Central Government. The NEC comprises of the Union Home Secretary as the Chairperson, and the Secretaries to the GOI in the Ministries/Departments of Agriculture, Atomic Energy, Defence, Drinking Water Supply, Environment and Forests, Finance (Expenditure), Health, Power, Rural Development, Science and Technology, Space, Telecommunications, Urban Development, Water Resources and the Chief of the Integrated Defence Staff of the Chiefs of Staff Committee as members. The other secretaries in the Ministry of External Affairs, Earth Sciences, Human Resource Development, Mines, Shipping, Road Transport and Highways.

8.4 The National Disaster Management System(NDMS)

After the Independence in India, the task of disaster management was undertaken by the Ministry of Agriculture, but after setting up of the national disaster management division (In February 2002), the Ministry of Home Affairs (MHA) was assigned the roles and responsibility of Disaster Management at the national level. The major change in the strategy is a paradigm shift from the culture of response to the culture of preparedness/mitigation

In the national system, various roles and responsibilities have been assigned at national and state level. The country has administrative machinery for management of disasters at national, state, and district as well as sub-district level. The subject of disaster management is handled by The Ministry of Home Affairs, Government of India that is the nodal Ministry for all matters concerning disaster management. The department of agriculture and cooperation is the nodal ministry for drought management. The Ministry of Environment and Forests deals with matters related to chemical disasters. A Central Relief Commissioner in the

Ministry of Home Affairs is the nodal officer to coordinate relief operations for natural disasters. A number of the ministries (Table 8.1) have been assigned a specific portfolio for disasters and their management.

Table 8.1: Ministries at national level and the task allocations

Type of Disaster	Nodal Ministry
Natural Disaster	Ministry of Home Affairs
Drought Relief	Ministry of Agriculture
Air Accidents	Ministry of Civil Aviation
Railway Accidents	Ministry of Railways
Chemical Disasters	Ministry of Environment
Biological Disasters	Ministry of Health
Nuclear Disasters	Department of Atomic Energy

Source: Disaster Management-The Development Perspective (2002-2007), Government of India, Ministry of Home Affairs, National Disaster Management Division.

The country has administrative machinery for management of disasters at national and state levels. The nodal ministry of Home Affairs has a special division known as National Disaster Management Division for coordination and looking after the natural disasters. Under the Crisis Management Group, different committees are formed at national and state levels to assigned the responsibility of each ministry and department in case of any disaster in the country.

Every ministry/department should act on the following points:

Allocate funds for measures for prevention of disaster, mitigation, capacity building and preparedness.

Provide assistance to the national authority and state government for

i) Drawing up mitigation, preparedness and response plans, capacity building, data collection and identification and training of personnel in relation to disaster management.

ii) Carrying out rescue and relief operations in the affected area.

iii) Assessing the damage from any disaster.

iv) Carrying out rehabilitation and reconstruction.

The framework of disaster management under the new vision is expected to undertake following arrangements:

- Encourage disaster mitigation and prevention
- Establish Institutional mechanisms to ensure coordinated efforts between various governments for disaster management
- Disaster management to be listed in concurrent list of the seventh schedule of the constitution

- Each of the states to draft the state disaster management act
- States to enunciate policy on disaster management
- Governments at various levels to ensure the formulation and implementation of disaster management codes
- Early warning systems to be developed at all levels, with the establishment of emergency operations centres
- Institutionalize knowledge and lessons learnt in the process of working on the national roadmap

The new institutional framework for disaster management in the country is shown in Figure 8.1.

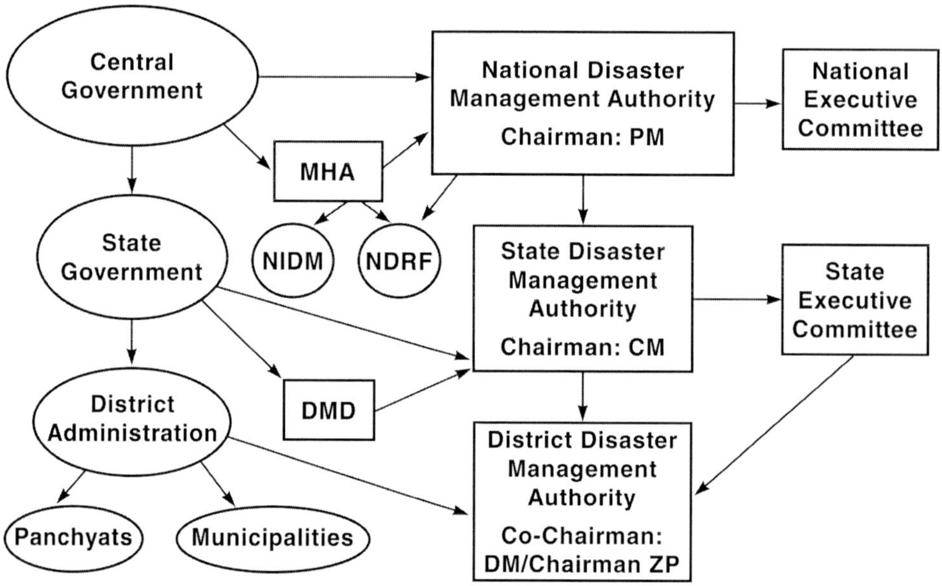

Figure 8.1 : Institutional framework of disaster management in India.

The Disaster Management Act 2005, lays down a three tier institutional structure for disaster management at the national, state and district levels in the form of NDMA, SDMA and DDMA. National Policy on Disaster Management (NPDM) has further specified the roles and responsibilities of various organizations for disaster response.

Most importantly, all the agencies and their functionaries must clearly understand their roles and responsibilities and they have to respond to disaster or threatening disaster situations.

All concerned Central Ministries, State Governments, District Authorities and other stakeholders will prepare a detailed standard operating procedure (SOPs) in consonance with this SOP, National Policy, and various Guidelines issued by NDMA.

These SOPs shall prescribe how various response and relief activities like evacuation, search and rescue, casualty and emergency health management, food, drinking water, sanitation and hygiene, clothing shelter, management of relief camps and restoration of essential services etc. have to be undertaken

The SOP encompasses the following five phases of disaster management for effective and efficient response to natural disasters.

(i) **Preparedness Phase –** This phase will include all necessary measures for planning, capacity building and other preparedness to be in a state of readiness to respond, in the event of a natural disaster. This Stage will also include development of Search and Rescue Teams, mobilization of resources and taking measures in terms of equipping, providing training, conducting mock drills/exercises etc.

(ii) **Early Warning Phase –** This phase will include all necessary measures to provide timely, qualitative and quantitative warnings to the disaster managers to enable them to take preemptive measures for preventing loss of life and reducing loss/damage to the property. On the occurrence of a natural disaster or imminent threat thereof, all the concerned agencies will be informed/ notified for initiating immediate necessary follow up action.

(iii) **Response Phase –** This phase will include all necessary measures to provide immediate help to the affected people by undertaking search, rescue and evacuation measures.

(iv) **Relief Phase –** This phase will include all necessary measures to provide immediate relief and help to the affected people in terms of their essential needs of food, drinking water, health and hygiene, clothing, shelter,

(v) **Restoration Stage –** This phase will include all necessary measures to stabilize the situation and restore the utilities.

8.5 Disaster Information Flow and Alerts

8.5.1 Standard Operating Procedures (SOPs)

The Ministry of Home Affairs (MHA) as the nodal Ministry is responsible for coordinating response and relief efforts with various ministries /departments of the Government of India, state government and district authorities. All the department and ministries have prepared the SOPs for handling disasters for which, they are earmarked as the lead ministry for disaster response, relief and mitigation. All central ministries, state government, district authorities and other stakeholders will prepare SOPs in consonance with the national and state plans.

Integrated Operation Centre (IOC) has been set up in the MHA to handle disaster situation on 24×7 basis. IOC is responsible for initiating incident alert messages, when a disaster is likely to occur or when it has actually taken place.

A standard SOP has been prepared for alerts of events of different types and identifies the situations, when alerts are to be sent by the IOC. Specific hazards have different categories of alerts. Accordingly, a uniform system has been devised by categorization each type of alert in stages – Yellow, Orange and Red (Annexure II).

8.5.2 Ministry of Home Affairs (MHA)

The Ministry of Home Affairs is the nodal agency at the national level for coordination of response and relief in the wake of disasters (except drought, pest attack and hailstorm). MHA will provide financial and logistic support to the State Governments, keeping in view, their resources, the severity of the natural disaster and the capacity of the State Governments to respond in a particular situation.

8.6 Different Institutional Mechanism

The institutional role and responsibility in case of disaster are fixed at different level and described below:

8.6.1 Institutional Mechanism at National Level

In the event of a disaster of a severe nature, national crisis management committee under cabinet secretary gives policy directions and guidelines to the crisis management group where national/international efforts are required.

Crisis management group in MHA reviews the situation in inter-ministerial meetings to coordinate various emergency support functions for the affected states.

Union Cabinet may set up a cabinet committee/task force for effective coordination of relief measures in the wake of calamities of severe nature.

8.6.2 Institutional Mechanism at State Level

A state level crisis management committee under the chairmanship of chief secretary is responsible for emergency management.

The committee comprises of concerned functionaries in various state departments and representatives of central organizations located in the state.

State relief/disaster management commissioner is the nodal officer for coordinating the activities for relief operations in the events of natural disasters.

8.6.3 Institutional Mechanism at District level

At district level, District Disaster Management Authority under the chairmanship of district magistrate is responsible for emergency management.

The different committees of various departments in the district and representative of the central government department in the district are included.

The District Magistrate ensures participation of district and state government in the response and recovery phase.

8.7 Committee for Disaster Management in India

In India, response to the disaster situation at all the levels of different committees has been made and discussed below:

Committee for Disaster Management in India

8.7.1 National Level

National Crisis Management Committee (NCMC)

This committee consists of representatives from ministries related to Agriculture, Defence, Finance, Science and Technology, Water Resources, Rural Development, Urban Development, Communications, Health, Petroleum, Power, Civil Supplies, Railways, Information and Broadcasting, Planning Commission, Surface Transport, Social Justice, Women and Child Development, Environment and Forest Food, Cabinet Secretary and Chief Secretary of the State.

At the National Level, the Command, Control and Coordination of the disaster response will be overseen by the National Crisis Management Committee (NCMC) under the Cabinet Secretary. NCMC will issue guidelines from time to time as required for an effective response to natural disasters. All Ministries/ Departments/Agencies at the national level shall comply with the instructions of NCMC.

National Executive Committee (NEC)

The NEC under the Union Home Secretary will 'coordinate response in the event of any threatening disaster situation or disaster'. NEC may give directions to the concerned Ministries/Departments of the Government of India, the State Governments and the State Authorities regarding measures to be taken by them in response to any specific threatening disaster situation or disaster.

8.7.2 Other Central Ministries/Departments

The other concerned central ministries/departments will render Emergency Support Functions (ESF), wherever their intervention and support are needed by any state governments in case of a disaster.

In case of a disaster or in any disastrous situation a set of sequence of actions is required at every level, *i.e.,* national, state and district levels. Following shall be the sequence of action at the national level.

NEC shall discharge the following functions:

1. On receipt of information either from State/District or from Early Warning Agencies or any other reliable sources, National Emergency Operation Centre (NEOC) of MHA shall be activated as per laid down protocol.
2. It shall send alerts/warning to all designated authorities.
3. Collect all relevant information and apprise the status to the designated authorities.
4. It shall activate ESFs if the situation warrants central assistance to States/UTs.
5. It shall prepare Daily Situation Report.

National Crisis Management Committee (NCMC)

1. On direction of the Cabinet Secretary, meeting of the NCMC shall be arranged.
2. NCMC will take stock of the situation.
3. It shall give necessary directions to NEC, MHA, other Central Ministries/ Departments/Agencies and State Governments.

National Executive Committee (NEC)

1. Joint Secretary, MHA shall take order from Union Home Secretary for convening the meeting of NEC
2. NEC shall assess the situation and give directions to the concerned Ministries/Departments of the Govt. of India, the State Governments and the State disaster management authorities regarding measures to be taken by them in response to any specific threatening disaster situation or disaster.
3. NEC shall coordinate response of various agencies.
4. It shall depute a team of officials to visit the affected states for on the spot assessment of the situation and coordinate with State Governments.
5. It shall mobilize resources and dispatch them to concerned States.
6. It shall monitor and review the situation regularly.

Role of the Nodal and Other Central Ministries/Departments

1. On activation of national ESF Plan, the concerned Ministry/Department shall depute the designated officials to NEC for coordination of response measures.
2. Ministries/Departments shall coordinate with their counterparts and make assessment of immediate needs in their respective sectors.
3. Ministries/Departments shall provide resources both in terms of men and material for assistance of the states on request of NEC.

8.7.3 State Level

State Crisis Management Committee (SCMC)

This committee consists of Relief Commissioner/Secretary, Department of Revenue, Collector/District Magistrate/Deputy Commissioner, District Level Coordination and Relief Committee.

State Executive Committee (SEC)

The primary responsibility of the state government is to respond to any disasters and provide relief to the affected people. SEC shall give directions to any department of the state government or any other authority or body in the state regarding actions to be taken in response to any disaster.

Department of Relief/Disaster Management shall be the nodal department for disaster management and Secretary of the Department/Relief Commissioner shall implement the decisions of the SEC pertaining to state level response to natural disasters.

Disaster response being a multi-agency function, other departments of the State Governments will provide emergency support in their relevant domains at the State/District levels. The responsibilities of SEC (State Executive Commitee) are as following:

1. Secretary (DM) shall convene the meeting of SEC.
2. SEC shall assess the situation and give directions to the concerned departments/agencies of the State Govt. and DDMAs concerned regarding measures to be taken by them in response to any specific threatening disaster situation or disaster.
3. SEC shall coordinate the response of various agencies.
4. SEC shall requisition NDRF or Armed Forces if the situation so demands.
5. SEC may depute a team of officials to visit the affected districts for on the spot assessment of the situation and supervise the response and relief measures.
6. SEC shall mobilize resources and dispatch them to concerned districts.
7. SEC shall monitor and review the situation on a regular basis.
8. SEC shall keep the NEC and NDMA informed of the situation.
9. SEC shall constantly evaluate their own capabilities to handle that situation and project the anticipated requirements for the central resources well in time.
10. SEC shall deploy state level Incident Command Team on the request of the DDMA as and when required.
11. In the event of the calamity of severe nature, the SEC will consider an appointment of senior officers, delegate powers and assign specific areas for timely and effective, efficient management of disasters.
12. SEC will also take necessary steps to pool the resources for better management of crisis situation. This includes the pooling of food grains funds and other resources available in the districts under various Government schemes.

However, these resources will be reimbursed to the respective schemes/departments once the situation becomes normal.

SEC shall discharge the following functions:

1. On receipt of information either from NEOC/DEOC or from Early Warning Agencies or any other reliable sources, State Emergency Operation Centre, shall be activated fully as per laid down protocol.
2. SEC shall issue alerts/warning to all designated authorities at the State level and Districts including for Public Information to AIR/Doordarshan/Press.
3. SEC shall send First Information Report to NEC, MHA and thereafter Daily Situation Report to situation normalizes.
4. SEC shall collect all relevant information and appraise the status to the designated decision making authorities.

5. It shall arrange meetings of SEC.
6. It shall activate ESFs of state if the situation so warrants.

State Disaster Management Authority (SDMA)

1. Meeting of the SDMA shall be convened on the direction of Chief Minister
2. SDMA will take stock of the situation.
3. SDMA shall give necessary directions to SEC, Deptt. of Disaster Management and other Departments/Agencies of the State Government and concerned DDMAs.
4. SDMA shall decide on Inter-State assistance and cooperation.

Role of Other Departments/Agencies

1. On activation of State ESF Plan, the concerned Department/Agency shall depute the designated officials to SEOC for coordination of response measures.
2. Departments/Agencies shall coordinate with their National counterparts and mobilize central assistance, such as specialized manpower, equipments, materials etc. to meet immediate needs in their respective sectors.
3. Departments/Agencies shall provide resources both in terms of men and material for assistance to the DDMAs.

8.7.4 District Level

District Disaster Management Authority (DDMA)

The administration head of the district is the chairman of DDMA and shall command and control for disaster response at the district level, in accordance with the policies/guidelines/instructions from the national and state levels. Depending on the nature of disaster and response, he will be the incident commander himself or delegate the responsibility to some other officer. Section 30(2)(xvi) of the Disaster Management Act stipulates that the DDMA under the chairmanship of the Collector or District Magistrate or Deputy Commissioner, as the case may be and shall 'coordinate to any threatening disaster situation or disaster'.

All the departments/agencies of the central and state governments in the district involved in response and relief will work following the directions given to the chairman of DDMA.

The lower administrative units of Districts, *viz.*, Subdivisions under the administrative control of a Sub-divisional Magistrate/Officer and Blocks and Tehsils under the administrative control of the Block Development Officers/Tehsildars will coordinate the functioning of the various departments in their respective jurisdiction.

At block levels, all SDO/SDM or BDO/Tehsildar be responsible for all response and relief work.

District Disaster Management Plans (DDMP)

The district administration is the focal point for implementation of all government plans and activities. Therefore, planning at the district level is crucial for efficient management of all disasters. The National Disaster Management Framework prepared by the Ministry of Home Affairs also emphasized on preparation of District Disaster Management Plans (DDMP).

The plan should also mention linkages with the NGOs and other organizations involved in disaster management at the district level. DDM plan should emphasize on mitigation (prevention and preparedness) and rehabilitation. Relief management should be given priority in the response plan.

- The structure of disaster management mechanism at the district level
- District Disaster Management Committee
- Disaster Management Teams
- Crisis Management Group
- Emergency Operation Centres and their operation.

Following shall be the sequence of action at the District level:

DEC shall discharge the following functions

1. On receipt of information either from NEC/SEC or from Early Warning agencies or field functionaries from Sub-divisions, Blocks, Tehsils or any other reliable sources and District Emergency Operation Centre shall be activated fully as per laid down protocol.
2. DEC shall issue alerts/warning to all designated authorities at the District level.
3. DEC shall send First Information Report to SEC and NEC, MHA and thereafter Daily Situation Report till situation normalizes.
4. DEC shall collect all relevant information and appraise the status to the designated decision making authorities.
5. DEC shall maintain all records and documents related to the response.
6. It shall activate ESFs of District if the situation so warrants.

Brief work of DDMA

1. DDMA shall assess the situation and give directions to the concerned Line Departments/Agencies at the District level regarding measures to be taken by them in response to any specific threatening disaster situation or disaster.
2. DDMA shall take such other actions as may be necessary for coordinated response to natural disasters. These may include the following:
    a. Assessing situations based on reports received from various sources and giving directions to different agencies for immediate response, relief and restoration of critical infrastructure.
    b. Reviewing the resources and capacities of different agencies to deal with

the situations and giving directions for pooling available manpower, equipment's and resources are available with different agencies for speedy and effective response.

c. Requisitioning assistance from NDRF/Armed Forces/other specialized agencies, if necessary.

d. Coordinating with civil society and Non-Governmental Organizations for supplementing the efforts of government agencies.

e. Monitoring and reviewing the situations on a regular basis.

8.8 Financial Mechanism and Activities

The entire Disaster Management Organisation has a financial mechanism in the Centre, known as Disaster Response Fund, which was earlier called Calamity Relief Fund. Now there is a permanent mechanism to make sure that even small things, which have to be done on ground, can be executed quickly. A national disaster response fund will be constituted and funded by central government and grant may be made by any person/institution. Funds allocated by ministry or departments for disaster management in their annual budget.

8.8.1 Financial Arrangements

Financial assistance from national and international agencies is very vital for post-disaster management for, relief, rehabilitation and reconstruction. These assistance are useful for providing food and medical facilities to the affected people and for other activities including establishment and operation relief camps over a long period of time. National and international assistance is used for providing building materials, technical equipment, facilitating agriculture re-establishment, feeding programmes, launching food for work programmes and extending support for livelihood opportunities. Reconstruction also needs outside assistance for planning and implementing interventions.

International assistance for disaster management can be obtained from the following agencies and organizations:

1. Foreign governments or their departments.
2. United Nations Organization and its organization like the Food and Agriculture.
3. Organization and United Nations Development Programme.
4. International aid agencies like the International Red Cross Society and World Vision.
5. International funding agencies like the World Bank and Asian Development Bank.
6. International technical agencies like the World Meteorological Organization.

National organizations and department agencies that can financial assistance in disaster management in an area include the following:

1. Central Government departments and organizations (National, state, district response and mitigation fund)
2. Non-governmental organizations.
3. Red Cross Society.

As per the annual reports, NDM division, Ministry of Agriculture, in the year 2001, about 788 lakh people were affected in the country by different disasters, with a damage/loss to property worth Indian Rupees 12000 Crores. As per the International database, in case of India, drought, earthquakes, floods and cyclones are the disasters that have been causing large scale damage to life and property in the last decade (Table 8.2).

Table 8.2 : Natural Disasters and their impact in India (1900 to 2004).

Event	No. of Events	Killed	Injured	Homeless	Affected	Total Affected	Damage US$ (000's)
Drought	21	4,250,430	0	0	1,391,841,000	1,391,841,000	1,898,721
Earthquake	24	60,396	206,561	1,935,700	24,966,300	27,108,561	5,124,900
Flood	157	50,860	1,268	8,934,230	654,151,850	663,087,348	11,118,841
Wind Storm	133	162,986	16,822	9,026,245	82,279,340	91,322,407	13,397,100

Questions for chapter evaluation

Descriptive questions

1. Write about disaster management policy and act. How these are different from each other?
2. What is national disaster management system in India and how it works?.
3. Write a note on different crisis management committee working at different levels.
4. Describe the different institutional mechanism at different level works in the country.
5. What are the different standard operating procedures in different phases of disaster?
6. Describe the financial mechanism and activities in India in case of a disastrous situation.
7. How does the district level committee works in case of a disaster in India?

Objective questions

Fill in the blanks

(i) The nodal agency for disaster management in India is ....

(ii) The nodal department looks after all the activities in case of nuclear disaster is ....

(iii) Disaster management is listed in seven schedules of the constitution of India in .... list

(iv) At national level, a fund for disaster financial management is made and is known as ....fund.

(v) The district magistrate will coordinate all the activities in the district in case of disaster by .... section of disaster management act.

Mention true and false against each of the questions given below.

(a) In a district, in case of disaster a central government agency is not under the command of district magistrate.

(b) Secretary of home ministry is to coordinate and response to all the ministries and central government departments in case of disaster in India.

(c) International assistance for disaster management can be obtained for reconstruction work after post disaster works.

(d) State relief commissioner is the nodal officer for coordinating the relief activity at district level.

(e) Disaster management policy was implemented in India in the year 2005.

CHAPTER 9

ROLE AND RESPONSE

In a disastrous situation, the role and responsibility of many institutions, organizations and public are required for its management. The role of government's, people, communities, non-government organization, media, police force and students in disaster management is to provide a central, coordinated plan of action to address the damage caused by such an event as well as the needs of the affected people.

The responsibility of every institutional or organization is to develop, review and assess effective disaster management practices. Help local government and the people for effective disaster management. Ensure local disaster management and disaster operations integrate with state and national disaster management.

The disaster planning team/groups should include representatives from different departments that will be mostly involved in policy development and the planning process. The response team will conduct the initial evaluation of facilities and systems following a disaster. When a disaster is identified, an appropriate team will be formed to assess and respond to the situation.

These include representatives from:

- **Government:** To make decisions involving required personnel and provide other financial, facilities to the people. Implementation of rules and regulations and provide relief and rehabilitation.
- **Police force:** To implement rules, coordinate and help for sustained recovery of the situation. Besides, provide peaceful working environment for other departments.
- **Technology:** Use of new tools and available technology for monitoring and creating awareness.
- **Human Resources:** To initiate communication procedures to general staff and contact staff necessary to fully evaluate disaster, to frame necessary internal and external communications
- **Media:** Work as a watchdog, stopping of rumours in society, and resource mobilization.
- **Industry:** To provide necessary requirement depending on the situations.
- **Other:** NGOs, self-help groups, community centres etc.

9.1 Roles

Government's role in disaster management is to provide a central, coordinated plan of action to address the damage caused by such an event as well as the needs of the affected people. Disaster management arrangements are characterized by and implemented through strong partnerships between government, government owned organization, NGOs, commerce and industry sectors and the local community.

The arrangements recognize and promote collaboration to ensure comprehensive disaster management through the effective coordination of disaster risk planning, services, information and resources.

9.1.1 Disaster Management Structures

These arrangements comprise several key **management** and **coordination** structures for achieving effective disaster management. Disaster management groups that operate at local, district, state and national levels are responsible for the planning, organization, coordination and implementation of all measures to mitigate/prevent, prepare for, respond to and recover from disaster events.

The management and coordination structures are:

- **Disaster management groups** that operate at local, district and state levels and are responsible for the planning, organisation, coordination and implementation of all measures to mitigate/prevent, prepare for respond to and recover from disaster events.
- **Coordination centres** at local, district and state levels that support disaster management groups in coordinating information, resources and services necessary for disaster operations.
- **Disaster management plans** developed to ensure appropriate disaster prevention, preparedness, response and recovery at local, district and state levels.
- **Functional lead agencies** through which the disaster management functions and responsibilities of the state government are managed and coordinated.
- **Hazard specific primary agencies**, responsible for the management and coordination of combating specific hazards.
- **Specific-purpose committees**, either permanent or temporary, established under the authority of disaster management groups for specific purposes relating to disaster management.

Role of different sections of society and organizations are required for effective implementation of the disaster management plans. Everyone role is very important and very much required for quick and mass scale recovery. The roles of different sections are described below:

9.1.2 Role of Local Government in a Disaster

The role of the government is to protect people against the natural calamities like floods, cyclones, famines, etc. The district level government has to provide every support to the people in a disaster and also,

- Have a disaster response capability
- Approve a local disaster management plan
- Pass the disaster information and coordinate with all other government and non-government agencies.

Local governments appoint local disaster management groups.

Their role is to:

- *Develop, review and assess effective disaster management practices*
- *Help local government to prepare a local disaster management plan*
- *ensure the community knows how to respond in a disaster*
- *Identify and coordinate disaster resources*
- *Manage local disaster operations*
- *Ensure local disaster management and disaster operations integrate with state disaster management.*

9.1.3 Role of Citizens

After the government, the role of citizens is very important during every phase of disaster management. The role of citizens may make a difference in their ability to cope with and recover from a disaster, as well as their ability to protect other household members and family possessions from avoidable losses.

Each individual should be prepared to take appropriate protective action if a disaster in going to occur. Understanding how the community would be warned, how to prepare the home, what supplies to stock, and how to check on family members following an emergency, are all important aspects of preparedness.

What can be done in advance

- All citizens should go through instructions in preparedness for all different disasters and then follow-through on being prepared. They should know the disaster plan and whom to turn to in an emergency.
- They should pay attention to emergency warnings and stay aware of developing situations.
- Citizens should be responsible for their own preparation and be able to carry on, if possible, after a disaster.
- They should make sure that they have extra supplies in case of a disaster: nonperishable food, water, candles, auxiliary power, blankets, first-aid kit, medications, generator, and money. They should know how to turn off the utilities in their homes.
- Learn how to use a fire extinguisher. Check them every year.
- Learn and continue to update skills in CPR and first aid.

- They should pay taxes to fund services that people need, donate blood, and pray.
- Public sensitization campaigns, education programmes through media as well as training on how to mitigate the impacts of disasters.
- Residents should continue to avoid living in high-risk zones and whenever they are issued with a relocation notice, they should always be willing to cooperate because it is in their interest.

What should be done after the disaster?

- Everyone should find a place to serve in a disaster. They should volunteer to help with disaster preparedness and relief.
- People need to stay positive, good-natured, kind, cooperative, and peaceful. They should be available to assist in any way that they are asked or directed to by government or NGOs.
- They should use their skills for assistance. For example, they can clear roads or assist firefighters and safety services, but should do this only as directed within the chain of command.
- Stay informed. Cooperate with agencies and comply with directions. For example, if there is an evacuation, they need to evacuate.
- Report injuries, looting, etc. Take pictures with cell phone.
- Citizens should organize, determine needs, can communicate to government and NGOs.
- After the initial disaster, they can collect money and raise funds to help those who are in need.
- Patients should be informed and assist, where they can.

9.1.4 Role of Armed Forces

Armed forces play a significant role in saving lives and rescue people. The National Disaster Management Policy acknowledges the role of the armed forces in disaster management and states that the armed forces are called, only when the coping capability of the civil administration is exhausted. The armed forces have shown their capabilities in communication, search and rescue, health and medical facilities, and transportation by air in many disasters. The armed forces are called out to assist and manage the situation.

The Indian armed forces respond to any disastrous situation. It is due to their technical competence, trained manpower, and logistical capabilities that they are always ready to rapidly undertake any kind of disaster-related rescue and relief operations. They are also located in most remote areas where natural calamities are frequent. For instance, when the tsunami hit the Indian coast on December 26, 2004, the Indian armed forces, co-coordinated by the Integrated Defense Staff (IDS), efficiently handled relief, rescue, and evacuation work under Operation Sea Wave, including extending aid. Generally, the armed forces respond to disasters as a part of their mandate to aid civil authorities during calamities. The roles played by the armed forces are numerous some are listed below:

- The fire at Burrabazar in Kolkata on January 12, 2008,
- The serial blasts at Bangalore and Ahmedabad in July 2008,
- The Mumbai attack of November 26, 2008,.
- The flash floods in Leh August 2010,
- The Sikkim earthquake of September 2011.
- The Amarnath landslide and flood during 2013.
- The flood in Jammu and Kashmir during 2015.

9.1.5 Role of Police Forces

There is no defined role of police in disaster management, but police has to reach the site of disaster to provide security and maintain law and order. They are also engaged to prevent commission of cognizable offences against life, property and public tranquillity.

The policy also envisages the potential of youth organisations such as the National Cadet Corps (NCC) to support all community-based initiatives and has suggested that they are trained in disaster management.

Advantages of police system in disaster management

- Well developed Communication System
- Wider Reach Every village covered
- Familiarity with Local Terrain and Area
- Awareness about People's Sensitivities

State Disaster Response Force (SDRF)

The SDRF is also be used for Community Capacity Building and Awareness Generation programmes within the state. Many states have made their own disaster response force like Odisha, Gujarat, Haryana, Jammu and Kashmir, Tripura, Arunachal Pradesh, Manipur, Tamil Nadu, Maharashtra. As per information available, 24 State/UTs have made their own SDRF. These SDRF are placed strategically at suitable locations well connected to the airport, rail heads and roads for their immediate deployment at the disaster sites.

National Disaster Response Force (NDRF)

It is a national level force, which is made by central government to be used for disaster relief and other rehabilitation work. Around 12 battalions of national disaster response force comprising more than 145 specialized teams have been trained for various types of natural and man-made disasters. Four battalions have been specially set up for handling radiological, nuclear, biological, and chemical disasters., but, considering the extent of India's hazardous environment and the escalating uneven trend in the occurrence of disasters, the present capability of the civil administration for combating disasters remain inadequate. The national disaster management author has made eight battalions taking personals from various central police forces. These battalions, with various skills, are located in different parts of the country based on the kind of disaster threats they are confronted with.

9.1.6 The Role of Non-governmental Organization (NGO)

No single organization is capable to manage a disaster and the ultimate responsibility rests with governments to protect its people against disaster. Government cannot carry out these responsibilities without cooperating with other groups or organization. An effective national disaster management system is largely based on coordination and cooperation of central, state, district and other organization.

NGOs play an important role in disaster response, mitigation, disaster reduction and recuse, coordination in different regions. Provide support like health, education, water supply and sanitation, shelter and infrastructure. NGOs are active in responding to various disasters in different phases.

NGOs can operate at grassroot level with communities and local organizations as partners, and take a participatory approach to development planning. NGOs have much more flexibility in their working and hence they better able to respond and adapt quickly and easily. NGOs often work with the poorest and the most vulnerable. NGOs can help government in identifying and reaching out to real victims and needy people.

Policy and Advocacy

A number of NGOs have been very active in advocating for better disaster risk reduction (DRR) policies and practices at the international level. Some played a key role in lobbying at the January 2005 World Conference on Disaster Reduction in Kobe, Japan, as well as in the follow-up of and implementation mechanisms for the Conference's outcomes. These efforts, made by NGOs recognized them for mainstream advocacy of DRR policies and development planning.

A significant number of the initiatives carried out at the community level are implemented by national NGOs and regional and sub-regional organizations. Still work is needed to identify NGOs that how they are helpful in practical terms.

Information Dissemination and Awareness

Disasters can be substantially reduced, if people are well informed and motivated towards a culture of disaster prevention and resilience. Several NGOs have played a key role in this area. Many of them have developed their own programmes and projects to educate local communities about the potential impact of natural hazards, and assist them in identifying possible ways to minimize their adverse consequences through collective and preventive action. Many of the NGOs have worked in India in disaster management and some are described below:

1. Gujarat School Safety Initiative, India

A major initiative was taken with help of a NGO (SEEDS) during devastating earthquake hit Gujarat in 2001. With a very high number of casualties among school children and many school buildings collapsed. Extensive rehabilitation programmes were required to do after the earthquake. The Gujarat State Disaster Management Authority (GSDMA) and the national NGO SEEDS took up the Gujarat School Safety Initiative, a first of its kind in this region to address the following issues

1. Understanding and preparedness amongst school children, teachers and parents to reduce disaster risk in schools and to be prepared to act in case of an emergency.
2. Disaster management appreciation amongst teachers, so that they are able to impart disaster education to children more effectively.

The Gujarat School Safety Initiative has piloted a methodology for school safety. The interest and involvement of the state government has expaned this pilot project into state and this project has benefited about 105,000 students across 175 schools in Gujarat state.

2. Aniruddha's Academy of Disaster Management (AADM)

Aniruddha's Academy of Disaster Management (AADM) is a NGO located in Mumbai, and its principal objective is to manage the disaster. The basic aim of AADM is to save life and property in the event of a disaster. It has successfully trained 60,000 citizens, the Disaster Management Volunteers (DMVs) to handle various disasters and disaster situations effectively. The AADM has built up a volunteer base, that assists the Government authorities during the disaster relief and rehabilitation works.

9.1.7 Role of Media

The media can create awareness among community about various types of disasters. It can also play a role in preparing the community by training them and making them aware about do's and don'ts during disasters. Only disaster preparation is not enough, awareness will greatly reduce the consequences of disasters in terms of human and economic loss. Electronic and print media are tools to disseminate information to a large number of people, quickly, effectively and efficiently. The suggestive, informative and analytical role of the media is useful for educating the community on disaster prevention, mitigation and rehabilitation. Media can play the dual role for imparting information for disaster management and also in analyzing the disasters. The electronic media, during Jammu and Kashmir earthquake in 2005, has demonstrated its potential in disaster management. Electronic media and information technology will play a major role not only the disaster response, but the overall disaster management strategy.

Media can be extremely effective in the following areas:

1. **Educational:** Educating people about the hazard, prevention before disaster and self-help during the disaster. Media can be helpful in providing, accurate and unbiased coverage during rehabilitation and post-disaster impacts.
2. **Guiding:** It can guide people in preparation of resource disaster calendar, resource mapping and community contingency plan.

3. **Critical:** It can evaluate the emergency plan and critically reviewed to any existing plan.
4. **Suggestive:** Media can suggest long term suggestions in the form of structured measure like enactment of certain legislation, adoption of code of conduct etc.

Other area where media can play effective role:

1. **Early warning to the people:** The media can help in providing an early warning to the people likely to be affected, or about the Do's and Don'ts during a disaster. The role of media in the preparedness of the community before the disaster, relief work during the disaster and rehabilitation after the disaster are very crucial. The early information by electronic media about the likelihood of disasters can save many humans beings and livestock and can also be minimized the loss to property to considerable extent.
2. **Watchdog on the disaster machinery:** The media can play the role of a watchdog, especially after the disaster, when rehabilitation work is going on. The rehabilitation work after the catastrophe is normally not carried out timely and there is response time delay of the relief machinery working after any disaster. The responsibility lies with the government to make available all the resources, but some time the system is little lethargic and respond little late. In those circumstances, media can play a vital role by being a watch dog and keeping the disaster machinery active.
3. **Appeal to the people:** The depiction of devastation and human misery through the media acts as an appeal to the people to come forward to render help in various ways. Generally, media has a responsive image in the public eye and this image can be utilized effectively to generate resources to help disaster management efforts.
4. **Prevent rumours and panic management:** During a disaster, with the breakdown of communications, rumour can have debilitating effect on the relief work. During this time, media can play a role in monitoring such rumours by relaying correct information about the measures being taken and monitoring them. Media can also help in establishing contacts, identifying the needy spots and focusing attention on them.
5. **Law and order controlling situation**: Media can keep a watch on anti-social elements, who try to take advantage of such disaster and emergency situations. They can report such matter and highlight such situations. They can also assist the law and order machinery in restoring peace and harmony in the affected community.
6. **Resource mobilization:** Media had played very effective role in raising resources for disaster victims in the past. During Tsunami disaster, Bhuj earthquake, the role of media was commendable; they had not only raised crore of rupees, but also supplied truck loads of relief materials–clothes food and medicine among affected people. They were able to convince people for larger participation by highlighting the contributions, providing tax-reliefs and quick and efficient mechanisms of donations.

7. **Use of established communication network:** During a disaster land-based telephone and telegraphy network is severely affected in India and during such situation media's network work (works on wireless mode), is very effective and useful for the administration. District administration who need to contact different agencies like Indian Air Force, Indian Navy, Army, Government officials and other departments for organization of relief have to rely on upon the network available with the media. The district administration take the support of media in making various announcements, releasing the list of casualty, lost people, do's and don't of the disaster etc.

Caution and Restrain necessary for media people

The media should be very careful during reporting of disasters. It should not create excitement and an atmosphere, which lead to panic. It is desired that a balanced approach is taken to report a situation. The people should be provided true and correct picture of disaster, its impact, the relief arrangements and coping mechanisms. They should avoid issuing warning signal unless it has been issued by the Govt.

Work list for media in disaster management

The rescue, relief and rehabilitation measures need an integrated and co-coordinated approach and for that purpose, all agencies, government and non-government, have to pool their resources together for efficient, expedient and effective work on all fronts. The depiction of devastation and of human misery through the media many times by itself acts as an appeal to the people to come forward to render help in various ways. Media have an important role, and a huge responsibility, in providing clear and accurate information to the public during a disaster. In addition to official emergency broadcasts, media have following works:

What happened?

Where it happened?

Who or what was affected?

What is being done?

Where it is safe to go?

When to distribute the items?

When it is safe to go back?

Where people can get help?

Where to get more information?

Who is ready to help?

Who is trying to take advantage?

Role of Social Media

Large portions of Indian population (230 million) are actively using social media platforms such as Whatsapp, Facebook, Instagram and Twitter. Around 70 per cent of the Indian population are online users and now connected to one or more social media platforms. People are not only keeping in touch with friends on social media, but also following and disseminating news among them. Social media can play a major role in helping people respond to emergencies and disaster. The social media platforms may be used :

- To disseminate information and receive user feedback
- To use social media more systematically as an emergency management tool.

9.1.8 The Role of Community Knowledge

There is too much dependence on Government to plan and implement disaster reduction and management works. To reduce the dependence on the government, there is a need to lay greater emphasis on formalized community participation in disaster mitigation and management measures. There is need to seek community participation in disaster preparedness and mitigation and their cooperation and assistance to minimize the damage from disasters. Community members play a vital role in reducing the impact of a disaster.

1. Role of Community

As it is the community, which is going to face the disaster, it is very important that community should be part of the complete disaster management process in all three phases–pre, during, post.

Community knows itself, its environment better. Many times, senior citizens are aware of incidences that have happened in the past, which others may not be aware of. One just needs to make community aware, and provide necessary inputs to identify hazards and risks.

Also in case of any disaster, the community itself acts as a first responder before the external help reaches to victims. So if the capacity building of community is done, community will be ready to face disaster and it will reduce the work-load on external agencies coming to help.

The community's role can be in all three phases– training, plan, early warning, communication, shelter, search and rescue, food arrangements etc. can be planned better with help of youths, women etc. from the community.

In a disaster situation, a large number of agencies, the Government and communities come together to help in coping with the after-effects of the disaster. In recent times, these agencies, also called civil-society have been collaborating even in times when there is no disaster, to put in place, for an effective Disaster Management Plan.

The community disaster team can follow the steps below to assess the risks and the resources that need to be addressed:

- Meet and discuss about the risks.
- Conduct visits to those sites exposed to risk.

- Clearly identify risks and locate them on a risk map
- Collect the information and have risk maps
- Strengthening inadequate shelters;
- Identifying places that can provide temporary protection against floods;
- Planning evacuation routes in the event of a fire.
- Make up of emergency service workers, local government leaders and other members of the community;
- List the resources available to the community in case of disaster;
- Propose actions to reduce the risks;
- Carry out efforts to reduce the risks;
- Risk assessment and their findings and conclusions.

2. Community action

In case of a disaster, the community takes part in determining its own evacuation routes and plans, its means of transportation, its next destination and how to access basic supplies. Every member of the community should be made familiar with these plans, so that evacuating people can go as smoothly as possible. Communities are frequently willing to have volunteers, who have positive reaction in case of disaster for the following types of activities:

- Warning people of danger;
- Searching for and rescuing victims;
- Giving food and shelter to the homeless;
- Constructing dams or dikes to eliminate flooding;
- Safely disposing of hazardous materials.
- Carry out drills and practice other basic disaster preparedness exercises in areas that can improve the victims' survival health and distributing food;
- Setting up temporary shelters and sanitation;
- Performing first aid;
- Transporting the sick and injured to hospitals.
- Reporting critical information by telephone, radio etc;
- Dealing with stray or dead animals.

3. Community Contingency Plan

Proposed plans of action in anticipation of a natural or human-made disaster. This involves:

(1) Identification of the potential threat, flood plains, history of drought, food shortages, or epidemics, earthquake etc.;

(2) Identification of likely impact of disaster, *e.g.*, number of people potentially affected, disruption of food or water supply, transportation

system, or communication channels, damage to property, roads, health facilities, duration of disaster and its effects;

(3) Identifying methods to mitigate the disasters such as shelter-belt plantations to break the intensity of cyclones along the coast;

(4) Anticipating and developing optimum response to such threats, *e.g.*, educate/alert population to potential risk, develop notification and evacuation plans, provide means of transporting people, food and medical supplies;

(5) Identification of existing resources, *e.g.*, areas, where shelters could be established, sources of food, water and medical supplies, communication and transportation systems;

(6) Conducting periodic 'mock drills', which are a simulation of a disaster to assess and improve the effectiveness of the disaster preparedness plan of a community or system.

9.1.9 Role of Students

When the students are trained well, they can help in rehabilitation and resettlement of victims at times of floods, earthquakes and drought etc. They will help the victims with basic needs and they will do whatever they can at times of any disasters.

9.1.10 Role of Technology

In recent times, technology has been employed to fast track disaster relief efforts. For instance, drones and robots have been used to locate survivors and transmit information to emergency teams. They have also been used to drop humanitarian aid.

9.2 Disaster Responsibilities

The government is responsible for ensuring the protection and welfare of the people during disasters or emergencies. It has the following responsibilities:

- Lay down policies on disaster management;
- Approves the national plan;
- Approve plans prepared by the Ministries or Departments of the Government in accordance with the National Plan;
- Lay down guidelines to be followed by the State Authorities in drawing up the State Plan;
- Lay down guidelines to be followed by the different Ministries or Departments of the Government of India for the purpose of integrating the measures for prevention of disaster or the mitigation of its effects in their development plans and projects;
- Coordinate the enforcement and implementation of the policy and plans for disaster management;
- Recommend provision of funds for mitigation;

- Provide such support to other countries affected by major disasters as may be determined by the Central Government;
- Take such other measures for the prevention of disaster or the mitigation, or preparedness and capacity building for dealing with threatening disaster situations or disasters as it may consider necessary;
- Lay down broad policies and guidelines for the functioning of the National Institute of Disaster Management.

9.2.1 Administrative Responsibilities

The basic responsibility for undertaking rescue, relief and rehabilitation measures in the event of natural disasters is that of the State Governments concerned. The role of the Central Government is supportive, in terms of physical and financial resources and complementary measures in sectors, such as transport, warning and inter-state movement of food grains. Relief Manuals and Codes are available for undertaking emergency operations.

In a disaster situation, a large number of agencies, the government and communities come together to help in coping with the after-effects of the disaster. A Disaster Management Plan (DAP) for a country is made up of several plans at various levels, such as the community, the block/taluka, the district and state integrating into a national plan. At the community level, it is called the community contingency.

A broad view of the administrative response at national, state and district levels is given below:

National level

Under the Indian federal system, disaster management is the responsibility of State Governments. However, there is a Crisis Management Group headed by the Cabinet Secretary and consisting of nodal ministries in charge of various types of disasters and supporting ministries. For natural disasters, the Ministry of Agriculture is the nodal Ministry and the other Ministries play a supportive role. In the event of a disaster, a multi-disciplinary Central Government team, at the invitation of the affected State, carries out disaster assessment and makes recommendation for assistance.

State Level

Disaster preparedness and response in the State is usually delegated to the Relief and Rehabilitation Department or the Department of Revenue. The Crisis Management Group at the State level is headed by the Chief Secretary of the Government, with participating of all the related agencies.

District Level

A District Level Co-ordination and Review Committee is constituted and is headed by the Collector as Chairman with participation of all other related agencies and departments.

9.2.2 Importance of Contingency Plan

Contingency planning aims to prepare an organization to respond well to an emergency and its potential humanitarian impact. Such planning is a management tool, involving all sectors, which can help ensure timely and effective provision of humanitarian aid to those most in need, when a disaster occurs.

The components of an emergency plan

- ✓ Planning
- ✓ Training
- ✓ Drills
- ✓ Education
- ✓ Technology
- ✓ Coordination
- ✓ Communication

9.2.3 Contingency Action Plan (CAP)

Disaster preparedness and contingency planning help to respond to disasters quickly and effectively. A contingency plan is a partial process of disaster preparedness planning. Contingency plans focus on certain events or risks, such as earthquakes or floods, at local, national or regional level.

Contingency planning refers to advance preparations that organizations make in case some unforeseen disastrous event happens. These type of plans may be made at national, state and district as well as local levels.

9.3 Disaster Preparedness

Disaster preparedness involves planning, identifying organizational resources, determining roles and responsibilities, developing policies and procedures and to respond timely and effectively to a disaster. Main aims of preparedness the actions taken to prepare for and reduce the effects of disasters. Means to predict and, where possible, prevent disasters, mitigate their impact on vulnerable populations, and respond to and effectively cope with their consequences.

Preparedness is a continuous cycle of planning, organizing, training, equipping, exercising, evaluating, and taking corrective action. These elements are the cornerstones of preparedness and focus on readiness to respond to all-hazards incidents and emergencies.

Every individuals should know the general disaster preparedness, *i.e.,* do's and don't under a disastrous situations. This may help anyone in facing a disaster and minimize the human suffering. A list of dos and don't for different hazard is given in appendix III.

9.3.1 Disaster Preparedness Tips

- Know what you'll face.
- Learn your area's evacuation routes and shelter locations.
- Know how you'll reconnect with people who matter.
- Sign up for emergency alerts and know how officials will communicate with you during a disaster.

- Learn what to do if you're caught away from home.
- Have a kit and know how to use it.

9.3.2 Mitigation and Preparedness

Mitigation includes prevention and risk reduction. Good evacuation plans, environmental planning and design standards are part and parcel of mitigation efforts. Preparedness refers to getting ready for facing a disaster.

Questions for chapter evaluation

Descriptive questions

1. What are the responsibilities of governments in the disaster management in India? List some responsibilities of the governments.
2. What is the role of a citizen in disaster management? List some of the role of a citizen during the disaster.
3. Describe the role of NGO in disaster management. Give some examples of NGO who worked for disaster management.
4. Write a note on role of media in disaster management and also mention some area where they may play their role.
5. What is the role of police force in disaster management? What is NDRF?
6. What is the role of social media in disaster management?
7. As a student how you can play a role in case of a disaster in India.
8. What is importance of a contingency plan? Describe the community contingency plan.
9. What is SDRF? What is its role in disaster management in a state?

Objective questions

Fill in the blanks

(i) .... per cent of Indian adult population is connected to the social media platform.

(ii) .... number of battalion of NDRF has been trained for various types of disaster.

(iii) The task of responding to any kind of disaster has been entrusted to the .... government.

(iv) The training and skill are common points of national and ...... government.

(v) The quick information dissemination of what and where about a disaster among public is a role of ....

Mention true and false against each of the questions given below.

(a) Media act as a watch dog on disaster machinery in case of occurrence of a disaster.

(b) Contingency plan is a usual plan framed to deal with a normal situation.

(c) Community action plan is not required to deal with disaster management.

(d) The police force is not very much required to deal with a disastrous situation.

(e) Citizen of India has no role to play in case of occurrence of a disaster.

CHAPTER 10

RESCUE AND REHABILITATION

Since disasters happen so rarely, management often ignores the disaster recovery planning process. This policy requires management to financially support and diligently attend to disaster contingency planning efforts. Rehabilitation, reconstruction and sustainable recovery refer to measures that help to restore the livelihoods, assets and production levels of emergency-affected communities.

Natural disasters and calamities cause widespread destruction, deaths and damage to property and infrastructure. Monitoring natural disasters like floods, earthquakes, volcanic eruptions, landslides, avalanches, cyclones, etc., are the main aims for managing the rescue and rehabilitation operations during and after such calamities.

The rescue, relief and rehabilitation activities are of importance in natural calamities and the activity mainly includes:

- To identify and locate the calamity hit regions to meet the contingencies arising out of the calamities and to arrange relief measures.
- To supervise and to provide funds for the rescue and relief operations.
- To monitor repair/restoration and rehabilitation work of the damaged properties and habitation.
- To undertake long-term measures by coordinating the activities of different NGOs, Government agencies, communities, citizens, etc. in order to minimize the impact of natural calamities and human casualties.
- To monitor hunger and starvation due to food scarcity on account of any disaster and provide food to the needy.

10.1 Rescue

Rescue comprises responsive operations that usually involve the saving of life or prevention of injury during an incident or dangerous situation. Its main role is to search for, assist and carry out rescue operations for the survivors by any means. Search and rescue are technical activities rendered by a group of specially trained personnel, who rescue and attend to the casualties under adverse conditions, where life is at threat. Search and rescue are organized in close cooperation with the community and a team approach.

Rescue departments provide urgent help in the event of an accident or the event of a potential hazard. The rescue may be of many types:

- Air-sea rescue.
- Cave rescue.
- Combat search and rescue.
- Confined space rescue.
- Helicopter rescue basket.
- Hostage rescue.
- Mine rescue.
- Mountain rescue.

10.1.1 Rescue Operations

Rescue comprises responsive operations that usually involve the saving of life or prevention of injury during an incident or dangerous situation. Rescue operations cover:

- Receiving emergency calls.
- Issuing warnings to the public.
- Preventing potential accidents or hazards.
- Protecting people, property and the environment against danger, and rescuing accident victims.
- Extinguishing fires and limiting damage.

10.2 Disasters and Rehabilitation

The term Disaster return to a whole range of distress situation, both individual and public. In the case of exogenous disasters, the first phase is threat and warning to the community of the disaster, the second phase will be emergency or crisis, when the community shares the suffering and impact; the third either of relief or deterioration and the fourth stage will be either recovery or rehabilitation. Very often, during the period of recovery and rehabilitation the poor and socially deprived not only lose their share of relief and resources, but can even be dispossessed of their homes and trade formerly located within the affected area. Often, the victims of disaster of exogenous category may also have to suffer a cumulative disaster – the endogenous being superimposed upon the exogenous. The initial responses in endogenous disaster are totally dissimilar to that of exogenous category. During the final phase, victims are weakened by starvation and roam aimlessly in search of food. At this stage, the victims appear to be mentally affected and are often the object of pity and horror, their survival centreed activities cause revulsion.

10.2.1 Rehabilitation

Rehabilitation consists of actions taken in the aftermath of a disaster to enable basic services to resume functioning, assist victims' self-help efforts to repair dwellings and community facilities, and to facilitate the revival of economic and agricultural activities.

The multi-sector rehabilitation programmes may be implemented to assist communities affected by the disaster. These programmes consist of disaster - resistant houses, education centres and have been set up drinking water supply systems, water harvesting structures, clinics for primary health care, and to establish other services to regain people livelihoods.

The main objectives of the rehabilitation programmes are:

- Reconstructing houses, a school and rural education centres damaged or destroyed by the disaster;
- Restoring livelihoods through the protection and management of local natural resources;
- Strengthening healthcare systems to improve access and the range and quality of essential services;
- Introducing water supply and sanitation systems;
- Empowering through training programmes to plan and manage equitable and participatory rehabilitation processes as well as initiate disaster preparedness of the community;
- Documenting and disseminating lessons learned.

Rehabilitation: The restoration of basic social functions.

Reconstruction: The full resumption of socio-economic activities plus preventive measures.

10.2.2 Resettlement and Rehabilitation

Rehabilitation and reconstruction include measures, which help increase the resilience of food systems in case of future disasters and emergencies. Rehabilitation includes the provision of temporary public utilities and housing as interim measures to assist long-term recovery. Reconstruction attempts to return communities to improved pre-disaster functioning.

Rehabilitation, reconstruction and sustainable recovery refer to measures that help restore the livelihoods, assets and production levels of emergency-affected communities. These measures rebuild essential infrastructure, institutions, services and restore the means of production destroyed or made non-operational by a disaster. Rehabilitation and reconstruction assistance is geared towards limiting the need for relief and allowing development activities to proceed. Rehabilitation and reconstruction include measures, which help increase the resilience of food systems in case of future disasters and emergencies. Major emphasis is on strengthening co-ordination of locally active emergency and development institutions and on encouraging the participation of the affected population in designing and implementing interventions to promote household food security and nutrition. Priority is given to the needs of food-insecure households and towards promoting sustainable livelihoods.

10.3 The Rehabilitation and Resettlement Bill (2007)

It provides for benefits and compensation to people displaced by land acquisition purchases or any other involuntary displacement.

> ***The Rehabilitation and Resettlement Bill, 2007***
>
> *Compensation offered to those displaced by natural disasters or by the acquisition of land by the government has often been described as inadequate. In this Bill, the government has attempted to specify a process by which people can be compensated as well as minimum levels of compensation that governments should pay.*

10.3.1 Highlights of the Bill

- The Rehabilitation and Resettlement Bill, 2007 provides for benefits and compensation to people displaced by land acquisition purchases or any other involuntary displacement. The Bill creates project-specific, state and national authorities to formulate, implement, and monitor the rehabilitation and resettlement process.
- For large scale displacement, the government shall conduct a social impact assessment. It shall appoint an Administrator for Rehabilitation and Resettlement, who is responsible for formulating, executing, and monitoring the rehabilitation and resettlement plan.
- The Bill outlines minimum benefits for displaced families and the criteria for eligibility. Benefits may include land, house, monetary compensation, skills training and preference for jobs.
- The Bill establishes the post of Ombudsman to address any grievances from the rehabilitation and resettlement process. Civil courts are barred from entertaining any suits related to this matter.

The International Strategy for Disaster Reduction (ISDR) defines recovery as the "decisions and actions were taken after a disaster with a view to restore or improve the pre-disaster living conditions of the stricken community, while encouraging and facilitating necessary adjustments to reduce disaster risk".

Recovery efforts following rescue and relief in any disaster can be classified into short term and long term. The short term activities for recovery are debris clearance, providing semi-permanent shelter and ensuring sanitation and restoring lifelines, while the long term activities involve building a safer and more sustainable livelihood.

10.4 Disaster Recovery Plan (DRP)

A disaster recovery plan (DRP) is a documented process or set of procedures to execute an organization's disaster recovery processes to recover and protect in the event of a disaster. It is "a comprehensive statement of consistent actions to be taken before, during and after a disaster".

10.4.1 Nature of Recovery

The damage caused by floods, earthquakes and cyclones is on a much larger scale than other disasters and recovery after these disasters pose a challenge. In disasters like drought, the relief phase is prolonged and since there is no damage to the infrastructure and property, the rehabilitation is confined to restoration of livelihoods, which can get subsumed in normal development programmes.

Recovery in case of epidemics is more in the form of sanitizing the locality against any future recurrence and may also involve counseling of the victims. Industrial disasters being quite varied in nature, the rehabilitation in major ones like the 'Bhopal Gas Tragedy' could involve rehabilitation efforts spanning over a generation of victims apart from restoring livelihoods and providing social and psychological assistance.

Rehabilitation following disasters such as landslides and avalanches is localized and is of a similar nature as in earthquakes, but on a smaller scale. Finding safer sites near such locations often pose challenges and resistance.

10.4.2 Guiding Principles for Post-Recovery

- Mainstreaming disaster risk reduction in recovery and development process,
- Improving coordination,
- Promoting participatory approaches and decentralizing planning and programming for recovery,
- Enhancing safety standards and integrating risk reduction in reconstruction and development,
- Improving the living conditions of the affected communities and sectors,
- Building local and national capacities for increased resilience, risk management and sustainable development,
- Taking advantage of ongoing initiatives,
- Gender sensibility,
- Demonstrative effects,
- Monitoring, evaluation and learning.

Questions for chapter evaluation

Descriptive questions

1. What is rescue and what are rescue operations?
2. How rescue and rehabilitation for disaster management are different?
3. Describe the different types of rescue operations.
4. What are the objectives of the rehabilitation?
5. Discuss about the disaster and rehabilitation in India.

6. What are recovery and its operations consists during disaster management?
7. Describe your role as a rescuer in case of occurrence of any manmade disaster.

Objective questions

Fill in the blanks

(i) The Rehabilitation and Resettlement bill is passed in the year ....

(ii) Rehabilitation, reconstruction and sustainable recovery refer to restore the ......

(iii) Search and rescue is a technical activity rendered by a group of specially ...... personnel.

(iv) ...... a comprehensive statement of consistent actions to be taken before, during and after a disaster

(v) Recovery efforts in any disaster can be classified into short term and ....

Mention true and false against each of the questions given below.

(a) Rehabilitation means restoration of basic social functions.

(b) Mainstreaming disaster risk reduction in recovery and development process.

(c) A disaster recovery plan (DRP) is a documented process or set of procedures to execute by any organization.

(d) The government will provide compensation for those who are displaced by any natural disasters.

(e) The rescue, relief and rehabilitation activities are not important in natural calamities.

CHAPTER 11

DISASTER EDUCATION

11.1 Disaster Education

Effective disaster management requires trained manpower to deal with complex situations effectively and speedily to reduce the impact of the disaster on human life and property. Capacity may also be described as capability and it may include physical, institutional, social or economic means as well as skilled personal. It is necessary to continuously undertake measures to build capacity amongst those who are handling disaster prevention, mitigation, preparedness, response, reconstruction and also creating awareness amongst people. Therefore, capacity building needs to encompass all resources available within a community, society and organisation to reduce the level of risk or the effects of a disaster.

The Government of India, Ministry of Human Resource Development (MoHRD) has emphasized the need for integrating disaster management in the existing education system in India. Education for disaster reduction cannot be a onetime affair, but should be reinforced time and again throughout one's life.

Knowledge management and education can help communities in hazard-prone areas to gain a better grasp of the ways to cope with risks. Knowledge and innovation, education, formal and informal are closely linked to disaster-reduction efforts. Disasters can strike at any time and it is the magnitude of the related impacts that will reflect the level of preparedness and "education" of the exposed country and community. It is now widely agreed that achieving disaster-resilience is essentially a process of using knowledge and learning at all levels.

11.1.1 Government Initiatives towards Disaster Education

World Conference on Disaster Reduction held in January 2005 concluded that "disasters can be substantially reduced if people are well informed and motivated towards a culture of disaster prevention and resilience". This can be achieved through collection, compilation and dissemination of relevant knowledge and information on hazards and vulnerabilities. It is a strong recommended that disaster response reduction knowledge should be introduced into school curricula and that other formal and informal channel should be used to disseminate this information among youth and children.

On the recommendation of 2nd Administrative Reforms Commission contained in its 3rd Report regarding the introduction of "Disaster Management" as a subject in 'Management and Public Administration'. The main teaching agencies University Grants Commission (UGC), Indian Council of Agricultural Research (ICAR), Indian Council of Medical Research (ICMR), Central Board of Secondary Education (CBSE) have incorporated the disaster management course in their teaching syllabus. The UGC framed the syllabus for an optional paper on Disaster Management at the Undergraduate level and syllabus for a short-term training course for UG teachers. The UGC approved the introduction of an optional paper on Disaster Management at the Undergraduate level across the universities/ colleges. The Commission further decided that Disaster Management be introduced as one of the topics in Orientation and Refresher Courses offered by the Academic Staff Colleges. The ICAR has inducted a Disaster management course in undergraduate and postgraduate course and taught in every agriculture colleges/universities in India. ICMR, Government of India, Ministry of Human Resource Development (MoHRD) in its tenth five year plan emphasized the need for integrating disaster management in the existing education system in India. In addition, the government of India launched a set of nation-wide disaster risk mitigation initiatives that addressed larger aspects of development in order to safeguard the developmental gains. One of the important initiatives includes disaster management in the curriculum of school and professional education has been recommended to the Boards.

The University Grants Commission (UGC) has finally issued a directive to all the 566 universities in India to introduce disaster management as an optional subject in all the undergraduate (UG) courses offered by them and their affiliated colleges. It has also circulated a model syllabus for the proposed subject. The directive is said to be the result of the high priority of the Union government for disaster mitigation through structured educational programmes and professional training. The four-member committee, appointed by the UGC to suggest measures for promoting disaster management education in the university system, has also recommended that disaster management should be a part of the orientation and refresher courses offered to teachers by the Academic Staff Colleges.

The Ministry of Home Affairs (MoHA), India launched a set of nation-wide disaster risk mitigation initiatives in its Tenth Five Year Plan to empower the younger generation to combat disasters. One of the important initiatives recommended to the education boards and institutions is the inclusion of disaster management in the curriculum of the school and professional education. The Central Board of Secondary Education (CBSE) has integrated a short course on Disaster Management in the school curriculum. The Board has introduced the subject on Disaster Management as a frontline curriculum in Social Science for classes VIII in the year 2003, for class IX in 2004 and for class X in 2005.

The state governments have been advised to take similar steps vis-à-vis their school boards. Several States Governments (SG) like Tamil Nadu, Sikkim, Delhi, Gujarat, Bihar, Maharashtra, Punjab and Odisha have introduced the Disaster

Management in their schools curriculum from Class VIII onwards. MoHRD is working with the Council of Board of School Education (COBSE) to make possible the insertion of disaster management in their public education in many school boards in India.

The Government of Andhra Pradesh issued orders to incorporate "Disaster Management" in the Social Sciences Curriculum of junior college and intermediate courses in the state. This is a step toward the dissemination of knowledge on disasters among the student community. The Government of Odisha in its resolution dated 4 March 2005 directed the state education authorities to "Make Disaster Management a part of the educational system and curricula".

> *UNDP describes 'capacity building' as the creation of an enabling environment with appropriate policy and legal frameworks, institutional development, including community participation, human resource development and strengthening of managerial systems. It adds that capacity building is a long-term, continuing process, in which all stakeholders participate.*

11.2 Components of the Capacity Development

The human resource development is imparted into its various modules for its outreach and awareness about disaster management. The following are the four types of education may be required:

1. Training

Training is important to provide knowledge, sharpening of skills, concepts, rules, or change of attitude about the disaster and its management. In India, many organization, institutes, and universities are providing training to the number of groups, organizations to prepare them to deal with a disastrous situation.

2. Education

Many degree, diploma and certificate courses on disaster management aspects are offered by a number of institutions in India. Similarly, disaster management is included in school curriculum for students at different levels of schooling in India.

3. Research

The national as well as many state-run institutions/universities are doing research in many fields of disaster to have more efficient management system.

4. Awareness

The national disaster management authority, national disaster management institute, state disaster management institutes, UGC and other training institutes organizing awareness programme for different sectors of personnel and public.

11.2.1 Capacity Development in India

In India, the national capacity development programme, corresponding with the intensity and extent of the hazard experienced in a different part of India, it needs to be evolved and implemented the capacity building at different regions at different levels. These capacity development programmes of resource enhancement should encompass all institutions, organizations and individuals that have a role in any part of the disaster management cycle.

11.2.2 Capacity Development Programme at Different Level

The capacity development programme is expending day by day in India. These programmes are running at different level *viz.*, organization level by MoHRD, MoHA, NADM at university level by the Indira Gandhi National Open University (IGNOU) and UGC are also running various courses in disaster management at undergraduate and postgraduate as well as doctoral level. Similarly, at institute level NIDM, DMI and different college are offering this course and Central Board of Secondary Education (CBSE) has also introduced the topic of Disaster Management as a part of the school curriculum in Social Science.

The training in the field of disaster management is required to make capable the people in handling emergency situations. Many colleges and universities in India offer courses in disaster management. These courses are offered at distinct levels, *viz.*, certificate, diploma, PG Diploma, degree and postgraduate degree etc.; research programs leading to the award of Doctor of Philosophy (Ph.D.) and post-doctoral research are also conducted in disaster management and allied fields.

11.3 Organization Level

11.3.1 Ministry of Human Resource Development (MoHRD)

As an effort to mainstream disaster management concerns into the education curricula, NDMA has taken the initiative in conjunction with the Ministry of Human Resource Development, to ensure that DM curriculum forms part of the course curriculum in undergraduate science and humanities courses and medical, engineering, architecture courses besides in school education. It will also focus on the training of teachers and faculties in disaster management. Sarva Siksha Abhiyaan (SSA) will also include the concerns of disaster resilience in their adult literacy curriculum.

11.3.2 National Disaster Management Authority (NDMA)

NDMA undertakes several initiatives in the process of capacity development. The SDRF battalions are trained in different institutes and colleges. The number of capacity building and awareness activities are carried out by the authority in India. Some of them are given below:

i) Mock drill activity

The authority undertakes mock exercises in vulnerable districts and industries on various types of natural and man-made disasters to help the state governments and district administration in reviewing the adequacy and efficacy of the state and district level. This type of exercises in schools also is conducted to empower the schools to face the disasters squarely before the specialist response is affected.

ii) Training activity

The training at grass level is very important, the authority has undertaken a capacity building programme for Panchayati Raj Institutions (PRI) in collaboration with IGNOU in many states. PRI are the local authorities at the grassroots levels and providing training is needed at all PRI of India.

iii) Awareness activity

Authority has launched several public awareness campaigns through electronic and print media. These programme objectives are to create knowledge for the target audience.

11.3.3 Ministry of Panchayati Raj and Rural Development (MoPRRD)

The Ministry of Panchayati Raj and Rural Development is issue directives about disaster management training to their representatives of the Panchayati Raj Institutions and local bodies. The Ministry of Panchayati Raj has also issued a letter during the year 2010 in this regard. Many States' efforts towards the training of Panchayati Raj functionaries and district level officials have been initiated in 11 states.

11.3.4 University Level

The Government has created and designated to offer degree, diploma and short term certificate course on disaster management to few institutes and universities. However, some of the private and government-aided college and university are also teaching the course on planning for disasters and emergency preparedness and providing training that offers short-term courses. The education in the field of disaster management started after 1992 and after that, a large number of universities initiated to offer different course in this field (Annexure-IV). The Indira Gandhi National Open University (IGNOU) was the first to offer a six-month certificate course in disaster management and also started the diploma course in disaster management in India. IGNOU also offers a comprehensive programme on community awareness in disaster preparedness. Most Central Universities (CU) of India have envisaged centre for disaster management under their school of environmental studies. Many other universities of India are now offering the disaster management courses and some of them are listed below:

- Guru Gobind Singh Indraprastha University, Delhi
- Sikkim Manipal University of Health, Medical and Technology
- Punjab University.
- Himachal Pradesh University
- Jamia Millia Islamia University, New Delhi.
- Anna University, Chennai

- IASE University, Rajasthan
- The Global Open University, Nagaland
- MS University, Vadodara
- Devi Ahilya University, Indore
- Sikkim Manipal University, Gangtok
- Nalanda Open University, Patna
- Alagappa University, Chennai
- Jawaharlal Nehru University, Delhi

11.3.5 Institute Level

Many institutes in India started the disaster management related course after 1995 and the first institute was the National Institute for Disaster Management (NDMI) and later many state governments and organization initiated the capacity building programme in India. Disaster Management Institute (DMI), Bhopal is one of the oldest state government institute in India to offer the capacity building in the field of disaster management in India.

i) NIDM

It is a premier institute for training and capacity development programs for managing natural disasters in India, on a national as well as regional basis. This institute was inaugurated by the union home minister on 11 August 2004. It is working for the government by providing policy advice and facilitating capacity building services including strategic learning, research, and training.

This institute is also providing e-learning courses to raise the level of awareness and preparedness as well as to enhance analytical skills.

ii) DMI

The government of Madhya Pradesh has established DMI in the year of 1987 in the backdrop of the Bhopal Gas Tragedy of the year 1984. The ultimate objective of the Institute is to 'prevent or mitigate the consequences of disasters and thereby benefit the general community'. This institute is conducting training and awareness programmes in disaster management and related subjects for the officials and executives of government departments, public and private sector. Besides this institute also offers post-graduate diploma and degree courses in the field of disaster management as well as conducts awareness and consultation work.

The many other institutes are working in the field of disaster management and offering the degree, diploma and certificate courses besides the training and some of them are listed below:

- The National Centre for Disaster Management (NCDM) set-up by the Indian Institute of Public Administration
- The Centre for Disaster Management set-up by Y.S. Chavan Academy of Development Administration

- Symbiosis Institute of Geoinformatics.
- TISS (Tata Institute of Social Sciences)
- PRT Institute of Post-graduate Environmental Education, Bhopal
- National Information Centre of Earthquake Engineering IIT Kanpur
- Disaster Mitigation and Management Centre, Dehradun
- Geological Survey of India, Kolkata
- Tata Institute of Social Sciences, Mumbai
- Centre for Disaster Management, Pune
- Indian Institute of Ecology and Environment (IIEE), New Delhi
- Indian Red Cross
- RedR
- Indian Institute of Public Administration

There are four training institutes of Central Para Military Forces (CPMFs), which impart training to NDRF battalions:

- National Industrial Security Academy NISA CISF, Hyderabad
- Basic Training Centre BTC ITBP Bhanu, Chandigarh
- Central Training College -II CRPF, Coimbatore
- BSF Insitute of Disaster Response BIDR ,Tekanpur, Gwalior

11.3.6 College Level

The first Disaster Management Training Institution of the country was founded on 29th April 1957 at Nagpur as the Central Emergency Relief Training Institute (CERTI) to support the emergency relief organization of the Government of India. This institute organized advanced and specialist training for revenue officials responsible for Disaster Relief Operations against any natural or manmade disaster. CERTI was renamed as National Civil Defense College (NCDC) on 1st April 1968. The other colleges are listed in Annexure–IV and some important and old colleges are given below:

- National Fire Service College (NFSC), Nagpur
- Christ College.

11.3.7 School Level

The Central Board of Secondary Education (CBSE) has integrated a short course on Disaster Management in the school curriculum. The Board has introduced the subject on Disaster Management as a frontline curriculum in Social Science for classes VIII in the year 2003, for class IX in 2004 and class X in 2005. Several States Governments like Tamilnadu, Sikkim, Delhi, Gujarat, Bihar, Maharashtra, Punjab and Odisha have introduced the disaster management in schools from Class VIII onwards.

11.4 Capacity Gaps in Disaster Management

Education for Disaster Management is a trans-disciplinary exercise aimed at developing knowledge, skill, and values at all levels. The Government of India in its X and XI Five year plan document has emphasized the need to enhance knowledge skills and values to reduce the impact of disasters.

The disaster management comprises of multi-sectoral issues and accordingly the capacity building is needed for all the sectors, which may play a pivotal role in managing the issue. NIDM reported that there exists a wide gap in the knowledge, skill, and attitude of the disaster managers for efficiently managing emergencies, particularly at the local and state level. It is suggested that to bridge this gap, it is important to have specific capacity development plans and strategies.

11.5 Benefits from Disaster Management Education

The capacity building will be equipping the individuals with the understanding, skills and access to information, the knowledge that enables them to perform effectively in any disastrous situation. Any organization after acquiring proper training will behave plans for efficient management the disaster situation and also minimize the losses so some extent. Organizations, institutions and agencies have to make legal framework development and regulatory changes to at all levels and in all sectors to enhance their capacities. After the implementation of disaster management act 2005 many organization, universities, institutes and college have started the various courses in the field of disaster management (Annexure-IV). Lists of available sources from where anyone understands and learns the skill towards capacity building are given in Annexure-V.

11.6 National Day for Disaster Reduction in India

To create awareness about disaster management, many society of India may celebrate 29th October as a day of National Day for Disaster Reduction (NDDR). The children shown in Figure 11.1 are creating awareness in their neighbourhood on the disasters.

Figure 11.1 : Children awareness programme on disaster management [*Source:* Nai Duniya]

Questions for chapter evaluation

Descriptive questions

1. Why education in the field of disaster management is required in India?
2. Write a note on the national institutes for disaster management in India.
3. What are the roles and responsibilities of national disaster management authority in disaster management education in India?
4. Write an about the objectives of a disaster management institute, Bhopal and list some institute of India offering courses in disaster management.
5. What are the roles of UGC in implementing the capacity building programme in the field of disaster management?
6. Write your view about the capacity building is disaster management.
7. List the name of institutes and organizations of your state involve is capacity building in the field of disaster management.

Objective questions

Fill in the blanks

(i) The oldest college offering training in the field of disaster management in India is ......

(ii) The DMI Bhopal is established in the year ......

(iii) The first the university to offer certificate course in disaster management is ....

(iv) The CBSE has started the disaster management course in the year ....

(v) The NIDM was inaugurated in the year ....

Mention true and false against each of the questions given below.

(a) Odisha government has added the disaster management courses its school education curriculum.

(b) UGC also offers the degree programme in disaster management.

(c) National Institute of Disaster Management is a private institute.

(d) The police force is not very much required to deal with a disastrous situation.

(e) BSF Institute of Disaster Response BIDR, Tekanpur, Gwalior is providing training to NDRF.

ANNEXURE I

THE DISASTER MANAGEMENT ACT, 2005

(ACT NO. 53 OF 2005)

An Act to provide for the effective management of disasters and matters connected therewith or incidental thereto.

Be it enacted by Parliament in the Fifty-sixth Year of the Republic of India as follows:—

1. Short title, extent and commencement.—(1) This Act may be called the Disaster Management Act, 2005.

(2) It extends to the whole of India.

(3) It shall come into force on such date1 as the Central Government may, by notification in the Official Gazette appoint and different dates* may be appointed for different provisions of this Act and for different States, and any reference to commencement in any provision of this Act in relation to any State shall be construed as a reference to the commencement of that provision in that State.

2. Definitions.—In this Act, unless the context otherwise requires,

(a) **"affected area"** means an area or part of the country affected by a disaster;

(b) **"capacity-building"** includes—

   (i) identification of existing resources and resources to be acquired or created;

   (ii) acquiring or creating resources identified under sub-clause (i);

   (iii) organisation and training of personnel and coordination of such training for effective management of disasters;

(c) **"Central Government"** means the Ministry or Department of the Government of India having administrative control of disaster management;

(d) **"disaster"** means a catastrophe, mishap, calamity or grave occurrence in any area, arising from natural or manmade causes, or by accident or negligence which results in substantial loss of life or human suffering or damage to, and destruction of, property, or damage to, or degradation of, environment, and is of such a nature or magnitude as to be beyond the coping capacity of the community of the affected area;

(e) **"disaster management"** means a continuous and integrated process of planning, organising, coordinating and implementing measures which are necessary or expedient for—

(i) prevention of danger or threat of any disaster;

(ii) mitigation or reduction of risk of any disaster or its severity or consequences;

(iii) capacity-building;

(iv) preparedness to deal with any disaster;

(v) prompt response to any threatening disaster situation or disaster;

(vi) assessing the severity or magnitude of effects of any disaster;

(vii) evacuation, rescue and relief;

(viii) rehabilitation and reconstruction;

(f) **"District Authority"** means the District Disaster Management Authority constituted under sub-section (1) of section 25;

(g) **"District Plan"** means the plan for disaster management for the district prepared under section 31;

(h) **"local authority"** includes panchayati raj institutions, municipalities, a district board, cantonment board, town planning authority or Zila Parishad or any other body or authority, by whatever name called, for the time being invested by law, for rendering essential services or, with the control and management of civic services, within a specified local area;

(i) **"mitigation"** means measures aimed at reducing the risk, impact or effects of a disaster or threatening disaster situation;

(j) **"National Authority"** means the National Disaster Management Authority established under sub-section (1) of section 3;

(k) **"National Executive Committee"** means the Executive Committee of the National Authority constituted under sub-section (1) of section 8;

(l) **"National Plan"** means the plan for disaster management for the whole of the country prepared under section 11;

(m) **"preparedness"** means the state of readiness to deal with a threatening disaster situation or disaster and the effects thereof;

(n) **"prescribed"** means prescribed by rules made under this Act;

(o) **"reconstruction"** means construction or restoration of any property after a disaster;

(p) **"resources"** includes manpower, services, materials and provisions;

(q) **"State Authority"** means the State Disaster Management Authority established under sub-section (1) of section 14 and includes the Disaster Management Authority for the Union territory constituted under that section;

(r) **"State Executive Committee"** means the Executive Committee of a State Authority constituted under sub-section (1) of section 20;

(s) **"State Government"** means the Department of Government of the State having administrative control of disaster management and includes Administrator of the Union territory appointed by the President under article 239 of the Constitution;

(t) **"State Plan"** means the plan for disaster management for the whole of the State prepared under section 23.

DISASTER MANAGEMENT ACT, 2005

The Act lays down institutional, legal, financial and coordination mechanisms at the National, State, District and Local levels. These institutions are not parallel structures and will work in close harmony. The new institutional framework is expected to usher in a paradigm shift in DMfrom relief-centric approach to a proactive regime that lays greater emphasis on preparedness, prevention and mitigation.

INSTITUTIONAL FRAMEWORK UNDER THE ACT

National Disaster Management Authority (NDMA)

The NDMA, as the apex body for disaster management, is headed by the Prime Minister and has the responsibility for laying down policies, plans and guidelines for DM and coordinating their enforcement and implementation for ensuring timely and effective response to disasters. The guidelines will assist the Central Ministries, Departments and States to formulate their respective DM plans. It will approve the National Disaster Management Plans and DM plans of the Central Ministries/Departments. It will take such other measures, as it may consider necessary, for the prevention of disasters, or mitigation, or preparedness and capacity building, for dealing with a threatening disaster situation or disaster. Central Ministries/Departments and State Governments will extend necessary cooperation and assistance to NDMA for carrying out its mandate. It will oversee the provision and application of funds for mitigation and preparedness measures. NDMA has the power to authorise the Departments or authorities concerned, to make emergency procurement of provisions or materials for rescue and relief in a threatening disaster situation or disaster. The general superintendence, direction and control of the National Disaster Response Force (NDRF) is vested in and will be exercised by the NDMA. The National Institute of Disaster Management (NIDM) works within the framework of broad policies and guidelines laid down by the NDMA.

The NDMA is mandated to deal with all types of disasters; natural or man-made.

Whereas, such other emergencies including those requiring close involvement of the security forces and/or intelligence agencies such as terrorism (counter-insurgency), law and order situations, serial bomb blasts, hijacking, air accidents, CBRN weapon systems, mine disasters, port and harbour emergencies, forest fires, oilfield fires and oil spills will continue to be handled by the extant mechanism, *i.e.*, National Crisis Management Committee (NCMC).

NDMA may, however, formulate guidelines and facilitate training and preparedness activities in respect of CBRN emergencies. Cross-cutting themes like medical preparedness, psycho-social care and trauma, community-based disaster, preparedness, information and communication technology, training, preparedness, awareness generation etc., for natural and man-made disasters will also engage the attention of NDMA in partnership with the stakeholders concerned.

Resources available with the disaster management authorities at all levels, which are capable of discharging emergency support functions, will be made available to the nodal Ministries/Agencies dealing with the emergencies at times of impending disaster(s).

[*Source:* NDMA, New Delhi report]

ANNEXURE II

CATEGORIZATION OF ALERTS

- For the purpose of dissemination of alerts a uniform system has been devised by categorizing each type of alert in stages - **Yellow, Orange and Red.**
- Alerts falling in **Yellow stage** will not be communicated to concerned central and state authority/administration through EMS.
- Alerts falling in **Orange stage** will be communicated to concerned central and state authority/administration with 12 hourly updates or when it is upgraded to the Red Stage, whichever is earlier.
- Alerts falling in **Red stage** will be communicated to concerned central and state authority/administration with 3 hourly updates or at more frequent intervals as warranted by the situation.
- Any changes in the category of alerts will be suitably integrated in the next message due as per the previous categorization.

CATEGORIZATION OF ALERTS

Specific hazards have different categories of alerts as indicated below.

Avalanches (Defence Research & Development Organization)

Category Description Stage

Red

All round : Very Unfavourable condition. Numerous large avalanches are likely from all possible avalanche slopes even on moderately steep terrain. Suspend all movements. Airborne avalanches likely.

Orange

High: Unfavourable condition. Triggering possible from all avalanche prone slopes even with low additional loads and reach the valley in large size. Suspend all movement. Airborne avalanches likely.

Yellow

Medium: Partly unfavourable condition. Triggering is possible on most avalanche-prone slopes with low additional loads and may reach the valley in medium size. Movement on slopes with extreme care. Valley movements with caution. Avoid steep slopes. Routes should be selected with care.

Yellow

Low: Generally favourable condition. Triggering is generally possible only with high additional loads and on very few extreme slopes. Only sluffs possible and reach valley in small sizes. Valley movement is safe. Movement on slopes with care

Tsunami (Department of Ocean Development)

Category Description Stage

Red

Tsunami Warning: Tsunami Alert Great Contains information about the earthquake and a tsunami evaluation message that a tsunami is expected.(eg. For earthquakes with > M6.5 occurring in the ocean within a depth of <100 kms, a tsunami warning will be issued for those areas falling within 60 minutes travel time from the earthquake source and if expected run-up is > 2m)

Orange

Tsunami Watch Moderate Contains information about the earthquake and a tsunami evaluation message that a tsunami is expected. (*e.g.*, For earthquakes with > M6.5 occurring in the ocean within a depth of <100 kms, a tsunami alert will be issued for those areas falling within 60 minutes travel time from the earthquake source and if expected run-up is between 0.5 to 2 m as well as for those areas falling above 60 minutes travel time from the earthquake source and if expected run-up is > 2 m)

Yellow

Slight Contains information about the earthquake and a tsunami evaluation message that a tsunami is expected. (*e.g.*, For earthquakes with > M6.5 occurring in the ocean within a depth of <100 kms,a tsunami watch will be issued for those areas falling within 60 minutes travel time from the earthquake source and if expected run up is < 0.5 m and for those areas falling above 60 minutes travel time from the earthquake source and if expected run-up is 0.5 to 2 m)

Landslides (Geological Survey of India)

Category Description Stage

Red

The slides that block busy pilgrimage routes during peak times resulting in hardships to thousands of pilgrims and sometimes resulting in loss of human life. Landslides, which result in blockade of courses of relatively large natural drainages. If the blockade is fairly large it could lead to formation of a very large reservoir of water behind it. Formation of a large landslide dam could result in sudden flooding of areas located upstream. Abrupt breaching of landslide dam would suddenly release enormous quantities of water in the downstream areas leading to flash floods that could result in loss of life and damage to property on large scale.

Orange

Landslides of large dimensions those are located over or in close vicinity of inhabited areas like urban settlements or fairly large rural settlements. Activity on these slides can result in loss of human lives, dwellings on large scale. These slides may also inflict heavy losses on urban infrastructure.

Orange

The landslides that may occur on the fringes of inhabited areas and result in limited loss of life and property. Landslides, which result in blockade of courses of relatively smaller natural drainages. If the blockade is of relatively smaller dimensions its impact would be of a lower order. Although a threat potential is there, it may not be immediate.

Yellow

Landslides which are fairly large and affect infrastructural installations like strategic and important highways and roads, rail routes and other civil installations like various appurtenant structures of hydroelectric and irrigation projects. The landslides that enter large water bodies like reservoirs of hydroelectric projects and could damage some of components of these projects.

Yellow

Landslides of small dimensions that occur away from habitations and do not affect either humans or their possessions. These may occur near infrastructural installations, agricultural and forestlands and may not affect them in a significant manner. These slides may include small incidents that block communication routes for short periods or do not affect the society in a significant manner.

Cyclone (India Meteorological Department)

Category Description Stage

Cyclone

Red

The location of the system at this stage may still be 300 km – 500 km away from the coast. Disaster Management Machinery is expected to be geared up fully during this phase. Post landfall outlook: During this phase warning messages are issued about 12 hours before actual landfall and are of a "very serious nature". At this stage, it is expected that the Disaster Management machinery is in full operational mode to face the impending disaster. All preparedness action should have been completed by this time. MHA would be closely monitoring steps taken by the concerned State Governments regarding evacuation and relief activities like food, sanitation etc. This phase is fit to be classified as "Great Danger" and all warning messages issued to MHA Control Room are required to be forwarded to senior officials of the PMO.

Orange

Cyclone Warning

These messages are issued 24 hours before commencement of bad weather and are of a "serious nature". During this stage, the system is monitored closely and the expected place & time of landfall and the districts along the coastal areas likely to be affected are clearly indicated in the warning messages.

Yellow

Alert Issued 48 hours before the commencement of bad weather when a system is located about 500 km or more away from the coast. The forecast may not contain information about landfall and hence it is still of informatory type but at the same time meant to trigger preparatory actions. During this stage, Disaster Managers plans on the course of action required to be initiated once the system moves closer to the coast.

Earthquake (India Meteorological Department)

Category	**Description Stage**	
Great	M > 7.0	**Red**
Moderate	5.0< M < 6.9	**Orange**
Slight	M< 5.0	**Yellow**

Flood (Central Water Commission)

Category Description Stage

Red

Unprecedented Flood (Water Level equal and above Highest Flood Level (HFL)

Orange

High Flood (Water Level less than Highest Flood Level but still within 0.50m. of the HFL)

Yellow

Moderate Flood (Water Level below 0.50m. less than HFL and above Danger Level)

Yellow

Low Flood (Water level between Warning Level and Danger Level)

Railways (Ministry of Railways)

Category Description Stage

Red

Major 100 or more deaths, or where additional assistance is sought by the Ministry of Railways

Orange

Medium 51 to 99 deaths.

Yellow

Minor 50 or more casualties (inclusive of deaths and injuries)

Forest Fire (Ministry of Environment & Forests)

Category Description Stage

Red

Major Fire

Large fire, which may result in substantial loss of human lives, massive environmental degradation or loss of wildlife.

Orange

Medium Fire

Where large forest area is under fire, which can be controlled by the State Government and no Central intervention is sought by the State Government.

Yellow

Ordinary fire and Localized fire can be controlled by the concerned territorial Conservator of Forests.

ANNEXURE III

DO'S AND DON'T DOS

FOR EARTHQUAKE

Do's

- Slip or drop to the ground; take cover yourself under a strong table/fixed furniture and hold on until the shaking stops. If there isn't a table or desk near you, cover your face and head with your arms and bend down in an inside doors or corner of the building.
- Stay in bed if you are there when the earthquake strikes. Hold on and protect your head with a pillow. In that case increasing shaking intensity, immediately move to the nearest safe place.
- Stay away from heavy furniture, machine, large panes of glass, shelves holding heavy objects, and anything that could fall, such as lighting fixtures or furniture.
- Minimize your movements to a few steps to a nearby safe place and if you are indoors, stay there until the shaking has stopped and you are sure exiting is safe.
- Use a doorway for shelter only if it is in close proximity to you and if you know it is a strongly
- Supported, load-bearing doorway.
- Stay inside until the shaking stops and it is safe to go outside.
- Cover your mouth with a handkerchief or clothing.
- Tap on a pipe or wall so rescuers can locate you.
- Use a whistle if one is available.
- Shout very loudly for help.
- Shouting can cause you to inhale dangerous amounts of dust.

Don't

- Don't move from one place to another during a severe earthquake.
- Do not use the elevators.
- Kitchens and garages tend to be the most dangerous. Also, know the safest place in each room.
- Do not light a match.
- Do not move about or kick up dust

FOR FIRE

Do's

- ➢ Cover your nose and mouth with a wet cloth and avoid inhaling through the mouth
- ➢ Get down, keep close to the ground and crawl towards the point of escape under the smoke.
- ➢ Inform the fire station (101) the moment fire breaks out.
- ➢ Don't run if your dress catches fire. Roll on the floor and try to put out the fire with blanket.
- ➢ Cling to the floor to escape from smoke and fire.
- ➢ Rescue team must know where to get buckets, water, sand, long bamboo, etc. to extinguish fire.
- ➢ Be aware that the electricity may go out or the sprinkler systems or fire alarms may turn on.

Don't

- ▪ Do not use lift/elevator.
- ▪ Do not take undue risk. Do not crowd the scene of fire.
- ▪ Do not tamper with any machinery during fire fighting.
- ▪ Leave them for authorized handling
- ▪ Do not argue or discuss at the scene of fire.
- ▪ Do not tamper with equipment, if you do not know its operation.

FOR CYCLONE

Do's

- ➢ Check houses, secure loose tiles by cementing wherever necessary, repair doors and windows.
- ➢ Keep some wooden boards ready so that glass windows can be boarded.
- ➢ Promptly demolish condemned buildings.
- ➢ Keep your television, radio and mobile-ready.
- ➢ Keep inform you and listen to latest weather warnings and advisories.
- ➢ Get away from low lying beaches or other locations which may be swept by high tides or storm waves.
- ➢ Get extra food, especially things which can be eaten without cooking or with very little preparation. Store extra drinking water in suitably covered vessel.
- ➢ Be sure that a window or door can be opened on the lee side of the house *i.e.*, the side opposite the one facing the wind.
- ➢ You should remain in shelters until informed by those in charge that you may return home.

- People should keep away from disaster areas unless you are required to assist.
- Relatives should be promptly informed about the safety of persons in the disaster area.

Don't

- Avoid being misled by rumours.
- Don't leave shelters until informed by the rescue personals.
- Don't leave the safer place during lull, however, minor repairs can be carried out.
- Don't touch the loose and dangling wire from lamp post, it may have electric current.

FOR LANDSLIDES

Do's

- Prepare tour to hilly region according to information given by weather department or news channel.
- Move away from landslide path or downstream valleys quickly without wasting time.
- Keep drains clean,
- Inspect drains for - litter, leaves, plastic bags, rubble etc.
- Keep the weep holes open.
- Grow more trees that can hold the soil through roots,
- Notice such signals and contact the nearest Tehsil or District Head Quarters.
- Ensure that toe of slope is not cut, remains protected, don't uproot trees unless re-vegetation is planned.
- Listen for unusual sounds such as trees cracking or boulders knocking together.
- Locate and go to shelters,
- Try to stay with your family and companions.
- Check for injured and trapped persons.
- Mark path of tracking so that you can't be lost in middle of the forest.

Don't

- Try to avoid construction and staying in vulnerable areas.
- Do not panic and loose energy by crying.
- Do not touch or walk over loose material and electrical wiring or pole.
- Do not build houses near steep slopes and near drainage path.
- Do not drink contaminated water directly from rivers, springs, wells but rain water if collected directly without is fine.
- Do not move an injured person without rendering first aid unless the casualty is in immediate danger.

FOR FLOOD

Do's

- ➢ Switch off electrical and gas appliances, and turn off services at the mains.
- ➢ Carry your emergency kit and let your friends and family
- ➢ Avoid contact with floodwater it may be contaminated with sewage, oil, chemicals etc.
- ➢ Walk in standing water with a pole or stick to ensure that you do not step into deep water, open manholes or ditches.
- ➢ Stay away from power lines electrical current can travel through water.
- ➢ Listen to the radio or television for updates and information.
- ➢ If the ceiling is wet shut off electricity.
- ➢ Use buckets, clean towels and mops to remove as much of the water from the afflicted rooms as possible.
- ➢ Place sheets of aluminium foil between furniture wet carpet.

Don't

- ▪ Walk through flowing water.
- ▪ Swim through fast flowing water.
- ▪ Drive through a flooded
- ▪ Eat any food that has come into contact with flood water.
- ▪ Reconnect your power supply until a qualified engineer and alert for gas leaks.
- ▪ Smoke or use candles, lanterns, or open flames.
- ▪ Scrub or brush mud and other deposits from materials.
- ▪ Stay away from ceilings those are sagging.
- ▪ Attempt to remove standing water using your vacuum cleaner.
- ▪ Remove standing water in a basement too fast.

After Floods

1. Drink chlorinated or boiled water.
2. Take clean and safe food
3. Sprinkle insecticides in the water ponds/stagnant water.
4. Please cooperate with disaster survey team by giving correct information.

FOR TSUNAMI

Do's

- ➢ Be familiar with the tsunami warning signals.
- ➢ Make sure all family members know how to respond to a tsunami. Pick an inland location that is elevated.
- ➢ Choose more than one evacuation route.

- Teach family members how and when to turn off gas, electricity, and water
- Children should be taught in advance about the evacuation plans
- Prepare emergency kit before hand. The emergency kit should contain Flashlight and extra batteries, battery-operated radio and extra batteries, First aid kit.
- Emergency food and water, Essential medicines etc
- Listen to a radio or television to get the latest emergency information, and be ready to evacuate if asked to do so.
- If you hear a tsunami warning, move at once to higher ground and stay there until local authorities say it is safe to return home.
- Move in an orderly, calm and safe manner to the evacuation site
- Stay tuned to a battery-operated radio for the latest emergency information.
- Help injured or trapped persons.
- Stay out of damaged buildings. Return home only when authorities say it is safe.
- Use a flashlight/torch when entering damaged buildings. Check for electrical shorts and live wires.
- Open windows and doors to help dry the building.
- Check food supplies and test drinking water.
- Fresh food that has come in contact with flood waters may be contaminated and should be thrown out.

Don't

- Do not use appliances or lights until an electrician has checked the electrical system.
- Not stay near by any beach. Never go down to the beach to watch a tsunami come in.
- Do not return home by following any rumors, follow only authorities advise it is safe to do so.
- Do not drink flooded water.
- Do not use the elevators.

FOR BIOLOGICAL DISASTER

Do's

- Close the doors and windows when a biological attack is imminent.
- Watch television, listen to radio, or check the Internet for official news and information including signs and symptoms of the disease, areas in danger, if medications or vaccinations are being distributed, and where you should seek medical attention if you become ill.
- Use common sense and practice good hygiene.

- However, if you notice of an unusual and suspicious substance nearby: Move away quickly.
- Cover your head and nose.
- Wash your hand with soap and water.
- Listen to the media for official instructions.
- Seek medical attention if you become sick.
- If you are exposed to a biological agent: Ultra-efficient filter masks can be used.
- Follow official instructions for disposal of contaminated items such as bag and cloths.
- Take bath with soap and put on clean clothes.
- Seek medical assistance. If required and advised, stay away from others or even quarantined.
- Exercise and meditate.
- Take your daily prescribed medicines regularly.
- Clean the frequently touched surfaces with disinfectant regularly.
- Monitor your health. If you develop any health difficulty immediately contact the nearest health care facility and follow the medical advice rendered.

Don't

- Don't go near affected/ sick people.
- Don't self-medicate.
- Don't shake hands or hug your friends and near ones.
- Do not go to the hospital for a routine checkup or follow up.
- Don't go to crowded places like parks, markets and religious places.
- Don't go out unless it is essential.
- Do n't assume that any illness is a result of the attack.
- Do n't pay attention to unofficial instructions via social media platforms.

ANNEXURE IV

List of Colleges, Institutes, Universities and other Organizations in India offering different courses in the field of Disaster Management

S.No.	Name of College / Institutes / Universities	Place
1.	Indian Academy of Fire and Safety	Nashik, Maharashtra
2.	Fire Safety and Disaster Management Institute (FSDMI)	Firozabad, Uttar Pradesh
3.	Zenith Safety Solutions	Tiruchirappalli,Tamilnadu
4.	Desh Bhagat University : School of Medical Sciences	Gobindgarh, Punjab
5.	Pondicherry university : Department of Ocean Studies and Marine Biology	Port Blair, Andaman and Nicobar Islands
6.	Odisha State Open University	Sambalpur, Odisha
7.	Imperial Institute of Fire and Safety	Kochi, Kerala
8.	Kerala University of Fisheries and Ocean Studies : School of Ocean Science and Technology	Ernakulam, Kerala
9.	Maya College of Disaster and Conflict Management	Dehradun, Uttarakhand
10.	Rajiv Gandhi University: Faculty of Environmental Science	Itanagar, Arunachal Pradesh
11.	Arunachal University of Studies : Faculty of Skill Development and Vocational Studies	Lohit, Arunachal Pradesh
12.	Amity Institute of Disaster Management	Noida,Uttar Pradesh
13.	National Institute of Fire and Safety Engineering	Una, Himachal Pradesh
14.	Institute of Land and Disaster Management (ILDM)	Thiruvananthapuram, Kerala
15.	Techo Glóbal University	Shillong, Meghalaya

S.No.	Name of College / Institutes / Universities	Place
16.	Fire Institute of Research and Engineering - IGNOU Community College	Sonipat, Haryana
17.	Haryana College of Fire and Safety Management (HCFSM)	Rohtak, Haryana
18.	Morni Community College	Panchkula, Haryana
19.	CEIT-IGNOU Community College	Srinagar, J & K
20.	Archbishop Powathil Assumption Community College (APACC)	Changanassery, Kerala
21.	North-Eastern Hill University: The School of Education	Shillong, Meghalaya
22.	Jamsetji Tata School of Disaster Studies	Mumbai, Maharashtra
23.	Asian Fire Engineering College (AFEC)	Nagpur, Maharashtra
24.	National Institute of Disaster Management	New Delhi, Delhi
25.	National Civil Defence College, Nagpur	Nagpur, Maharashtra
26.	Guru Gobind Singh Indraprastha University	Delhi, Delhi
27.	Sikkim Manipal University of Health Medical and Technology Sciences	Gangtok, Sikkim
28.	Punjab University	Chandigarh, Panjab
29.	Himachal Pradesh University	Simla, Himachal Pradesh
30.	Jamia Millia Islamia University	New Delhi, Delhi
31.	Anna University	Chennai, Tamilnadu
32.	IASE University	Sardarshahr, Rajasthan
33.	The Global Open University	Dimapur, Nagaland
34.	MS University	Vadodara, Gujrat
35.	Devi Ahilya University	Indore, Madhya Pradesh
36.	Sikkim Manipal University	Gangtok, Arunachal Pradesh
37.	Nalanda Open University	Patna, Bihar
38.	Alagappa University	Chennai, Tamilnadu
39.	Jawaharlal Nehru University	Delhi, Delhi
40.	Disaster Management Institute	Bhopal, Madhya Pradesh
41.	National Industrial Security Academy NISA CISF	Hyderabad, Telangana

S.No.	Name of College / Institutes / Universities	Place
42.	Basic Training Centre BTC ITBP Bhanu,	Chandigarh, Haryana
43.	Central Training College -II CRPF	Coimbatore, Tamilnadu
44.	BSF Institute of Disaster Response BIDR Tekanpur	Gwalior, Madhya Pradesh
45.	Indian Red Cross Society	New Delhi, Delhi
46.	Y.S. Chavan Academy of Development Administration -The Centre for Disaster Management	Pune, Maharashtra
47.	Symbiosis Institute of Geoinformatics	Pune, Maharashtra
48.	TISS (Tata Institute of Social Sciences)	Mumbai, Maharashtra
49.	PRT Institute of Post-graduate Environmental Education	Bhopal, Madhya Pradesh
50.	National Information Centre of Earthquake Engineering IIT Kanpur	Kanpur, Uttar Pradesh
51.	Disaster Mitigation and Management Centre	Dehradun, Uttarakhand
52.	Geological Survey of India	Kolkata, West Bengal
53.	Tata Institute of Social Sciences	Mumbai, Maharashtra
54.	Centre for Disaster Management	Pune, Maharashtra
55.	Indian Institute of Ecology and Environment (IIEE)	New Delhi, Delhi
56.	Indian Institute of Public Administration	New Delhi, Delhi
57.	Indira Gandhi National Open University (IGNOU)	Delhi, Delhi
58.	Central Board of Secondary Education (CBSE)	Delhi, Delhi
59.	Christ College	Irinjalakuda, Kerala
60.	National Fire Service College (NFSC)	Nagpur, Maharashtra
61.	Delhi University	Delhi, Delhi

ANNEXURE V

READING MATERIALS ON DISASTER MANAGEMENT

S.No.	Name of the Book	Name of the Author	Publisher
1.	Academy for Disaster Management Planning & Training	ADEPT	Kilpauk, 2009
2.	Guiding the United Nations 2007, New York, NY	Annan K	Infobase Publishing, 2007
3.	Knowledge Management and education for Disaster Reduction, France	Rouhban B.	UNESCO, 2010
4.	Disaster Management	Sharma V.K.	National Centre for Disaster Management, 2001
5.	Disaster Management	M. M. Sulphey	PHI Learning Pvt. Ltd, 2016
6.	At risk: natural hazards, people's vulnerability and disasters	Wisner, Ben; Blaikie, Piers; Cannon, Terry; Davis, Ian	CRC Press, Taylor & Francis, 2003
7.	Disaster Management and Preparedness	Larry R. Collins	CRC Press , Taylor & Francis, 2000
8.	Disaster Management Handbook	Jack Pinkowski	CRC Press, Taylor & Francis, 2008
9.	Development and Disaster Management– A Study of the Northeastern States of India	Singh, A., Punia, M., Haran, N.P., Singh, T.B. (Eds.)	Palgrave Macmillan, 2018
10.	Natural Hazards and Disaster Management: Vulnerability and Mitigation	R. B. Singh	Rawat Publications, 2006
11.	Disaster Management: Global Challenges and Local Solutions	Rajib Shaw	Universities Press (India) Private Ltd, 2009
12.	The Disaster Preparedness Handbook: A Guide for Families	Arthur T. Bradley	Skyhorse publishing, China, 2011

S.No.	Name of the Book	Name of the Author	Publisher
13.	Handbook of Disaster Risk Reduction & Management	Christian N Madu, Chu-Hua Kuei, *et al.*	World Scientific, 2017
14.	Natural Hazards and Disaster Management	B. C. Jat	MD Publication Pvt Ltd., 2009
15.	U.S. Emergency Management in the 21st Century From Disaster to Catastrophe	Claire B. Rubin and Susan L. Cutter	CRC Press , Taylor & Francis, 2019
16.	Handbook of Disaster Research	Rodriguez, Havidan, Quarantelli, Enrico, Dynes, Russell (Eds.)	Springer, 2007
17.	Disaster Management: A Reader	S. Rajagopal Sridhar K. Chari	National Institute of Advanced Studies, 2005
18.	Disaster Mitigation: Experiences and Reflections	Pardeep Sahni, Alka Dhameja, Uma Medury	PHI Learning Pvt. Ltd., 2001
19.	Introduction to International Disaster Management	Damon P. Coppola	Butterworth-Heinemann, 2011
20.	Coping with catastrophe :A Handbook of Disaster Management	Hodgkinson PE and Stewart	Routledge, 1991
21.	Disaster Management	Gupta, H.K.	University Press, 2003
22.	Disaster management notes	online	smartworld.asia
23.	Manual for Drought Management	Anonymous	Department of Agriculture, Cooperation & Farmers Welfare Ministry of Agriculture & Farmers Welfare Government of India, 2016
24.	Disaster-induced migration: assessing the impact of extreme weather events on livelihoods	Krishnamurthy, P. K.,	Environmental, Hazards, 2012
25.	Southeast Asia tsunami December 2004	Shangle, Robert D	American Products Publishing Company, 2005

S.No.	Name of the Book	Name of the Author	Publisher
26.	Disaster Reduction: Living in harmony with nature	Kurowa, Julio	Quebecor World Peru, S.A.
27.	Human Development Report 2005: Internation co-operation at a cross roads: Aid, trade and security in an unequal world	Watkins, K	UNDP, 2005
28.	Disaster risk reduction 1994–2004	ISDR	ISDR
29.	The other side of paradise	WMO	
30.	Sri Lanka: Post tsunami environmental assessments	UNEP	United Nations Environmental Programme, 2005
31.	Tsunami	United Nations	UN/ISDR, 2006
32.	Tsunami: Progress in predicting disaster prevention and warning	Tsuchiya Yoshito & Shoto Nobuo	Kluwer Academic Publishers
33.	Advances in natuaral and technological hazards research: Tsunami research at the end of a critical decade	Hebebstreit Gerhard T	Kluwer Academic Publishers
34.	Status of coral reefs in tsunami affected communities, 2005	Wilkinson Clive, Souter Davis & Goldberg Jeremy	Australian Institute of Marine Science, 2006
35.	Tsunami hazard: A practical guide for tsunami hazard reduction	Bernard EN	Kluwer Academic Publishers, 1991
36.	Thirty years of natural disaster: 1974–2003: the numbers	Guha-Sapir D, Hargit D & Hoyois P.	UCL Presses Universitaires, 2004
37.	Disaster risk reduction: a development concern	White, Philip, *et al.*	DFID
38.	World disaster report: Focus on information in disasters	Jonathan, Walter	Internation Federation of Red Cross and Red Crescent Societies
39.	Know risk	Jeggle, Terry	Tudor Rose for United Nations, ISDR, 2005

S.No.	Name of the Book	Name of the Author	Publisher
40.	Mitigation of natural hazards and disasters: International perspectives	C Emdad Hague	Springer, 2005
41.	Living with risk: A global revision of disaster reduction intiatives (Vol 1)	United Nations	ISDR, 2004
42.	Living with risk: A global revision of disaster reduction inititiaves (Vol 2)	United Nations	ISDR, 2004
43.	Mainstreaming disaster risk reduction: A tool for development organisations	La Trobe Sara & Davis Ian (Prof)	Tearfund, 2005
44.	Status of coral reefs in tsunami affected communities, 2005	Wilkinson Clive, Souter Davis & Goldberg Jeremy	Australian Institute of Marine Science, 2006
45.	Advances in natural and technologival hazards research: Perspectives on tsunami hazards reduction, observations, theory and planning	Heberstreit Gerhard T	Kluwer Academic Publishers
46.	Geological hazards– their assessments, avoidance and mitigation	Bell, Fred G	Spon Press, 1999
47.	Catastrophe modeling: a new approach to management crisis	Gross, Patricia & Kunrheuter Howard	Springer, 2005
48.	Rising from the ashes: Development strategies in times of disaster	Anderson Mary B & Woodrow Peter J	Intermediate Technological Publishers, 1998
49.	Natural hazards and disasters	Donald Hyndman & Hyndman David	Thomson Brooks/ Cole
50.	Natural Hazards: Earth's processes as hazards disasters and catastrophe	Blodgett Robert & Keller Edwards	Pearson Prentice Hall, 2006
51.	Advances in natural technological hazards research: International perspectives on natural disasters Occurrences, mitigation and consequences	De Chano LM, Lidstone J & Stoltman Joseph M	Kluwer Academic Publishers, 2004

S.No.	Name of the Book	Name of the Author	Publisher
52.	Global environmental risk	Kasperson Jeanne X & Kasperson Robert E	Earthscan, 2001
53.	Mapping vulnerability: Disasters, Development & People	Bankoff G, Frerks G & Hilhorst, D	Earthscan Publications, 2004
54.	Reducing the risk of disasters – Helping to achieve sustainable poverty reduction in a vulnerable world: A DFID policy paper	DFID	DFID
55.	Components of Risk: A Comparative Glossary	Thywissen K	United Nations University UNU-EHS (Institute for Environment and Human Security)
56.	Disaster Management in India	Anonymous	Ministry of Home Affairs Government of India, 2011
57.	Emergency Handbook	Sridhar Chiruvolu, Pankaj Kumar Singh, Abhinav Walia,	Centre for Disaster Management Lal Bahadur Shastri National Academy of Administration, Mussoorie, 2018